90 07

D1756594

Lecture Notes on Coastal and Estuarine Studies

Managing Editors:
Richard T. Barber   Christopher N.K. Mooers
Malcolm J. Bowman   Bernt Zeitzschel

4

Howard R. Gordon
André Y. Morel

Remote Assessment of Ocean Color for Interpretation of Satellite Visible Imagery
A Review

Springer-Verlag
New York   Berlin   Heidelberg   Tokyo   1983

Authors

Howard R. Gordon
Department of Physics
University of Miami
Coral Gables, FL 33124, USA

André Y. Morel
Laboratoire de Physique et Chimie
Université Pierre et Marie Curie
06230 Villefranche-sur-Mer FR

Library of Congress Cataloging in Publication Data
Gordon, Howard R.
   Remote assessment of ocean color for interpretation
of satellite visible imagery.
   Bibliography: p
   1. Optical oceanography—Remote sensing.   2. Colors—
Analysis.   I. Morel, André Y.   II. Title.
GC180.G67   1983   551.46'01   83-16984

Printed and bound by Halliday Lithograph, West Hanover, Massachusetts.
Printed in the United States of America.

9 8 7 6 5 4 3 2 1

ISBN 0-387-90923-0 Springer-Verlag New York Berlin Heidelberg Tokyo
ISBN 3-540-90923-0 Springer-Verlag Berlin Heidelberg New York Tokyo

Acknowledgments

This work is a contribution to the research encouraged by the IAPSO Working Group on Optical Oceanography, in particular its third term of reference ('Examination of ocean optical properties with their application to other aspects of oceanography, including physical oceanography, ocean dynamics, heat absorption and climatology, marine biology, and sedimentology.'). Due to the important amount of oceanographic data used in this study which have been acquired over a long period, all of the support received cannot be acknowledged. By restricting ourselves to the latest, we wish to acknowledge the Centre National de la Recherche Scientifique (under contracts ERA 278 and GRECO 034), the Centre National d'Etudes Spatiales (under contracts 80-81/250), the Centre National d'Exploitation des Oceans (under contracts 79/2084, 79/2010, and 81/203), the National Aeronautics and Space Administration (under contract NAS 5-22963 and grant NAGW-273), and the National Atmospheric and Oceanic Administration (under contract NA-79-SAC-00741).

We also wish to acknowledge the support and effort of the following individuals: L. Prieur for providing the unpublished data used in Figure 5b; D.K. Clark for providing Figure 13; J.W. Brown, O.B. Brown, and R.H. Evans for processing the imagery presented in Appendix II; and D.K. Clark and D.L. Ball for additional processing of Figures AII-3 through AII-5.

CONTENTS

I. INTRODUCTION

Since the pioneering work of Clarke et al. (1970)
it has been known that chlorophyll a (or, more
generally, pigments) contained in phytoplankton in
near-surface waters produced systematic variations in
the color of the ocean which could be observed from
aircraft. As a direct result of this work, NASA
developed the Coastal Zone Color Scanner (CZCS), which
was launched on Nimbus-G (now Nimbus-7) in October
1978. (A short description of the CZCS is provided in
Appendix I.)

Shortly before launch, at the IUCRM Colloquium on
Passive Radiometry of the Ocean (June 1978), a working
group on water color measurements was formed to assess
water color remote sensing at that time. A report
(Morel and Gordon, 1980) was prepared which summarized
the state-of-the-art of the algorithms for atmospheric
correction, and phytoplankton pigment and seston
retrieval, and which included recommendations
concerning the design of next generation sensors.

The water color session of the COSPAR/SCOR/IUCRM
Symposium 'Oceanography from Space' held in Venice
(May 1980, i.e., in the post-launch period) provided
the opportunity for a reassessment of the
state-of-the-art after having gained some experience
in the analysis of the initial CZCS imagery. Such an
assessment is the purpose of this review paper, which
will begin with an outline of the basic physics of
water color remote sensing and the fundamentals of
atmospheric corrections. The present state of the
constituent retrieval and atmospheric correction
algorithms will then be critically assessed,
considering primarily the papers presented at the
Symposium, and also some new material. Samples of

imagery are presented and the initial comparisons with experiments are discussed. Finally, an appendix (Appendix II) is provided which summarizes the major developments in water color remote sensing which have taken place during the preparation of the report, and prior to the IAMAP Assembly in Hamburg (August 1981).

II. PHYSICS OF OCEAN COLOR REMOTE SENSING

A typical water color remote sensing situation is depicted in Fig. 1, which shows the satellite's radiometer aiming toward a spot on the sea surface (marked PIXEL), hence measuring the radiance[*] $L_T(\theta,\emptyset)$ emerging from the Earth's atmosphere. The goal of this measurement is to determine the concentration of the various constituents of the water (e.g., phytoplankton, total suspended material, yellow substances, etc.). The radiance $L_T(\theta,\emptyset)$ consists of photons which have been multiply scattered from the atmosphere, ocean, and sea surface, and thus depends in a complex manner on the optical properties of the atmosphere (and their distribution with altitude) and

---

[*] A radiometer consists of a flat detector of surface area A, capable of measuring the radiant power ϕ falling on it in a spectral band $\Delta\lambda$ centered on the wavelength λ. When the detector is located at a position specified by the vector $\underset{\sim}{r}$, is aimed in a direction specified by the unit vector $-\underset{\sim}{\xi}$, and is viewing the universe with a solid angle of acceptance $\Delta\Omega$, it measures the radiance traveling in the direction $\underset{\sim}{\xi}$ defined by

$$L(\underset{\sim}{r},\underset{\sim}{\xi},\lambda) \equiv \phi \; / \; A \; \Delta\lambda \; \Delta\Omega.$$

In the text the direction of $\underset{\sim}{\xi}$ is specified by the angles θ and $\emptyset$. When λ is omitted from the argument of L it is to be understood that the statement or relationship is true for all wavelengths.

3

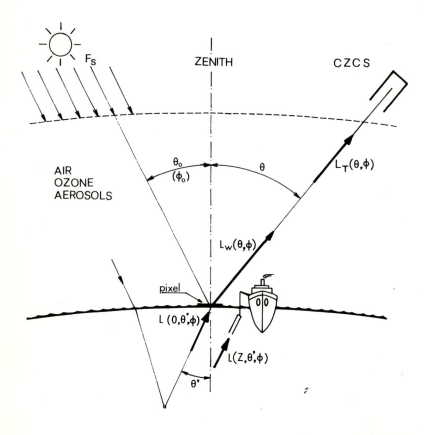

Figure 1. Schematic geometry and nomenclature for upwelling radiances.

on the optical properties of the ocean (and their distribution with depth). Through a series of Monte Carlo simulations (Gordon, 1976) of radiative transfer in an ocean-atmosphere system characterized by realistic and vertical distributions of all pertinent optical properties in the atmosphere, and a representative set of optical properties for the ocean (assumed homogeneous), three simplifying facts emerge. First, the vertical structure of the atmosphere has only minor influence on $L_T(\theta,\emptyset)$. Next, the ocean and atmosphere may be decoupled for the purpose of computing $L_T(\theta,\emptyset)$. Finally, the effect of ocean constituents on $L_T(\theta,\emptyset)$ can be adequately simulated by placing a hypothetical lambertian reflector of albedo R just beneath the sea surface (i.e., at depth z=0). The required subsurface albedo R is the irradiance ratio or irradiance reflectance defined by

$$R \equiv E_u(0)/E_d(0),$$

where the upward and downward irradiances, E_u and E_d respectively, are defined by

$$E_u(0) = \int_0^{2\pi} \int_0^{\pi/2} |\cos\theta'| L(0,\theta',\emptyset') \sin\theta' d\theta' d\emptyset',$$

$$E_d(0) = \int_0^{2\pi} \int_{\pi/2}^{\pi} |\cos\theta'| L(0,\theta',\emptyset') \sin\theta' d\theta' d\emptyset',$$

and $L(0,\theta',\emptyset')$ is the radiance a submerged observer would measure (in the ocean to be simulated) with a radiometer aimed at an angle θ' from the nadir and an azimuth angle $\emptyset'$ relative to the sun (see Fig. 1). Mathematically,

(1) $L_T(\theta,\emptyset) = L_1(\theta,\emptyset) + R\,L_2(\theta,\emptyset)/(1 - rR),$

in which $L_1(\theta,\emptyset)$ is the contribution to $L_T(\theta,\emptyset)$ from all photons which never penetrated the ocean (but may have reflected from the surface), $L_2(\theta,\emptyset)$ is the radiance which results from photons which penetrated the surface and interacted once with the lambertian surface for the case R=1, and r is a constant equal to about 0.5 (see, e.g., Austin, 1974 and Gordon, 1976). The factor $1/(1-rR)$ in the second term accounts, to all orders, for photons which are internally reflected in the ocean; since R very rarely exceeds 0.1, this factor is of minor importance.

The interpretation of $L_T(\theta,\emptyset)$ in terms of ocean constituents thus requires understanding the relationship between R and the constituents, while the retrieval of the constituents from $L_T(\theta,\emptyset)$ requires an accurate estimate of $L_1(\theta,\emptyset)$, which typically accounts for more than 90% of $L_T(\theta,\emptyset)$.

A. Irradiance Ratio and Upwelling Subsurface Radiance

The physics of 'ocean color' is well understood. The ocean color is physically described by the spectral values of reflectance $R(\lambda)$. The first theoretical approaches were made by using the two-flow (Schuster's) method, as an approximate solution of the radiative transfer problem (Gamburtsev, 1924, cited in Kozlyaninov, 1972; Duntley, 1942; Joseph, 1950; and Kozlyaninov and Pelevin, 1965). R was found to be related to the absorption coefficient, a, and backscattering coefficient, b_b, through (Joseph, 1950)

$$R = \frac{(1 + 2b_b/a)^{1/2} - 1}{(1 + 2b_b/a)^{1/2} + 1},$$

where b_b is linked to the volume scattering function $\beta(\gamma)$ according to

$$b_b = 2\pi \int_{\pi/2}^{\pi} \beta(\gamma) \sin\gamma \, d\gamma,$$

and to the scattering coefficient, b, through $b_b = \bar{b}_b b$, where $\bar{b}_b$ is the dimensionless backscattering ratio, and b is given by

$$b = 2\pi \int_0^{\pi} \beta(\gamma) \sin\gamma \, d\gamma.$$

Symbols and definitions used here follow the terminology adopted by IAPSO (International Association for Physical Sciences of the Ocean, see Morel and Smith, 1982). The Gamburtsev–Duntley expression for R is

$$R = \frac{b_b^*/a^*}{1 + (b_b^*/a^*) + (1 + 2b_b^*/a^*)^{1/2}},$$

where a^* and b_b^* are the absorption and backscattering coefficients for the diffuse light stream, called 'hybrid' optical properties by Preisendorfer (1961). They are related to the 'true' or 'inherent' properties of the ocean by

$$a^* = a/\bar{\mu}_d \quad \text{and} \quad b_b^* \cong b_b/\bar{\mu}_d.$$

$\bar{\mu}_d$ is the average cosine for the upper hemisphere which characterizes the downward radiance distribution and is obtained by forming the ratio

$$\bar{\mu}_d = E_d/\overset{o}{E}_d,$$

where $\overset{o}{E}_d$ is the downward scalar irradiance

$$\overset{o}{E}_d = \int_0^{2\pi} \int_{\pi/2}^{\pi} L(z,\theta',\emptyset')\sin\theta'd\theta'd\emptyset'.$$

Note that the average cosine for the entire radiance field, $\bar{\mu}$, which will be used later (III.B), is defined as $\bar{\mu} = E/\overset{o}{E}$, where E is the magnitude of the (downward) vector irradiance defined to be $(E_d - E_u)\underset{\sim}{z}$, with $\underset{\sim}{z}$ a unit vector in the nadir direction, and $\overset{o}{E}$ is the scalar irradiance

$$\overset{o}{E} = \overset{o}{E}_d + \overset{o}{E}_u = \int_0^{2\pi} \int_0^{\pi} L(z,\theta',\emptyset')\sin\theta'd\theta'd\emptyset'.$$

In the limit $b_b \ll a$, which is valid for nearly all oceanic and coastal waters and for the spectral domain considered, and also with the realistic approximation $b_b/a = b_b^*/a^*$, the above formulae simply reduce to

$$R \cong \frac{1}{2} b_b/a.$$

The main value of this formulation is the demonstration that, at least to first order, R is governed by the ratio of backscattering to absorption.

Full computations of radiative transfer in the ocean have now been carried out for various inherent properties of the medium and various illumination conditions. Gordon et al. (1975) used a Monte Carlo technique and their computations of R were fitted to polynomial expansion

(2)
$$R = \sum_{1}^{3} r_n X^n$$

with

$$X = \frac{b_b/a}{1 + b_b/a}.$$

In this expansion the first term is predominant, with r_1 having a value of 0.32 for solar illumination with the sun near the zenith and 0.37 for totally diffuse illumination. Note that unlike a* and b_b^*, a and b_b are rigorously summable over the constituents of the ocean, i.e.,

(3)
$$a = a_w + \sum (a)_i$$

and

(4)
$$b_b = (b_b)_w + \sum (b_b)_i,$$

where w refers to water, and i to the i-th constituent. The ocean constituents influence R through their effect on (a) and (b_b). In general (Morel and Prieur, 1977),

$$(a)_i = (a)_i^0 C_i$$

(5)

$$(b_b)_i = (b_b)_i^0 C_i,$$

where $(a)_i^0$ and and $(b_b)_i^0$ are the absorption and back scattering coefficients per unit concentration of the i-th constituent (i.e., the specific absorption and backscattering coefficients for the i-th constituent).

By the use of the successive scattering order method (Prieur and Morel, 1975; Prieur, 1976) an alternative expression was obtained,

(6) $R = 0.33 \ (b_b/a) \ (1 + \Delta).$

The second order term Δ was studied as a function of phase function and of the radiance distribution within the submarine light field. In any case, Δ remains weak (typically less than ±5%) and for practical purpose may be neglected. The usefulness of this very simple relationship has been demonstrated in many papers (Morel and Prieur, 1977; Smith and Baker, 1978b), and recently confirmed by Kirk (1981). At least for remote sensing studies, the use of more complicated relationships or attempts to refine them through new calculations of radiation transfer are not needed until a better knowledge of the absorbing and scattering properties of the materials present in the sea is reached.

All of the computations described above were carried out for a homogeneous ocean (i.e., a and $\beta(\gamma)$ independent of depth), whereas in practice the ocean is usually highly stratified. Based on the Monte Carlo simulations of radiative transfer on strongly stratified media, Gordon and Clark (1980a) have concluded that R is still approximately given by Eq. 2 if X is replaced by

$$\overline{X} = \int_0^{\tau 90} g(\tau) X(\tau) \, d\tau \Big/ \int_0^{\tau 90} g(\tau) \, d\tau,$$

with

$$g(\tau) = \exp[-2 \int_0^{\tau} K(\tau') \, d\tau'/c(\tau')],$$

where K_d is the attenuation coefficient for downwelling irradiance, defined as

$$K_d(z,\lambda) = d(Ln[E_d(z,\lambda)])/dz,$$

c is the beam attenuation coefficient (with $c = a + b$), and τ is the optical depth which is related to the true depth z through

$$\tau = \int_0^z c(z) \, dz.$$

The upper limit on the integrals above is the optical depth at which the downwelling irradiance falls to $1/e$ of its value just beneath the surface. For a homogeneous ocean, the depth z_{90} corresponding to τ_{90} is simply $1/K_d$ and is called the 'penetration depth.' Only 10% of the contributions to R results from photons which have reached depths greater than z_{90} (Gordon and McCluney, 1975).

When the optical properties of the ocean are largely governed by the concentration (C) of a single constituent, then Gordon and Clark (1980a) have shown that the 'effective' concentration (the concentration which would yield the same R in a homogeneous ocean) is

(7) $$C_f = \int_0^{z_{90}} C(z) f(z,\lambda) dz \Big/ \int_0^{z_{90}} f(z,\lambda) dz,$$

where

$$f(z,\lambda) = \exp\left[-2 \int_0^z K_d(z',\lambda) dz'\right].$$

Thus a constituent at a depth z_{90} would have to be about an order of magnitude more concentrated than at the surface to have the same effect on R.

Interpretation of Remotely Sensed Ocean Color

In the case of a remote sensor, the quantity actually measured is a radiance L. After atmospheric correction the quantity to be interpreted is $L_w(\theta,\emptyset)$, the water-leaving radiance. (Austin (1974) calls L_w the inherent sea surface radiance; however, we shall adopt the term 'water-leaving radiance' to avoid the confusion which may arise because radiance is an apparent optical property rather than an inherent optical property, according to standard terminology.) This is the radiance just above the surface which emerges from the ocean (see Figure 1). For a flat surface it is related to the upwelling radiance just beneath the surface $L(\theta,\emptyset)$ by

$$(8) \qquad L_w(\theta,\emptyset) = L(\theta,\emptyset)[1 - \rho(\theta',\theta)]/m^2,$$

where m is the refractive index of water, $\rho(\theta',\theta)$ the Fresnel reflectivity (subsurface) for an incident angle θ', and $\sin\theta' = (1/m)\sin\theta$. Since m varies only slightly with wavelength, $L_w(\theta,\emptyset)$ and $L(\theta,\emptyset)$ have essentially the same spectral composition.

To the approximation that the ocean is a diffuse Lambertian reflector (beneath the surface) the radiance $L(\theta',\emptyset')$ has a constant value L_u (for downward looking paths), and

$$E_u = \pi L_u,$$

or

$$L_u = (E_d(0)/\pi)R.$$

From radiance distribution measurements (Tyler, 1960; Smith, 1974) and also from theory (Prieur, 1976), it appears that the above assumption is an oversimplification. According to Austin (1974), if

L_u is the radiance in the nadir direction, π must be replaced by a factor closer to 5. This is principally due to the fact that $L(\theta',\emptyset')$ increases rapidly with θ' for $\theta' > \theta_c'$ the critical angle (~ 48°). For $\theta' < \theta_c'$ the assumption that $L(\theta',\emptyset')$ is constant in a given situation appears to be well satisfied, and hence

$$L_u = (E_u(0)/Q)R.$$

According to theoretical predictions, Q depends slightly on wavelength λ. If L_u is used in absolute units, this factor must be taken into consideration. If radiances are used through ratios as $L_u(\lambda_1)/L_u(\lambda_2)$ for the two wavelengths λ_1 and λ_2, the spectral change of the above mentioned factor should be considered, but has not been, due to the lack of experimental data. At present, the influence of this quantity on the algorithms discussed below is felt to be small.

When the sea surface is roughened by the wind one expects the relationship between $L_w(\theta,\emptyset)$ and $L(\theta',\emptyset')$ to be more complex than Eq. 9. However, Austin (1974) has shown that when $L(\theta',\emptyset')$ is completely diffuse the dependence of the surface transmittance $(L_w(\theta,\emptyset)/L(\theta',\emptyset'))$ on wind speed is weak, and Eq. 9 can be applied with an error of less than 10% for wind speeds less than 10 m/s and view angles less than about 50 degrees. Even though $L(\theta',\emptyset')$ is typically not totally diffuse, the fact that it is essentially constant for $\theta' < \theta'c$ implies that this conclusion is still valid. Again, if the radiances are used through ratios, the ratio of water-leaving radiances will be the same as the ratio of subsurface radiances.

Along with being well understood theoretically, ocean color has also been well documented experimentally. Since 1970, considerable data (see Table 1) have been acquired concerning spectral values $E_u(\lambda)$, $R(\lambda)$ and more recently, $L_u(\lambda)$. Some of them remain unpublished. The waters studied vary from the 'desert' blue waters (e.g., Sargasso Sea) to those rich in phytoplankton, in suspended sediments, in dissolved yellow substance, or influenced by diverse combinations of these various components. As a matter of fact, algorithms now in use in CZCS data interpretation, or others more advanced in use for airborne measurements, originate from analytical or statistical studies of this 'radiometric sea-truth' data bank generated by ship-bound oceanographers.

Interpretation of Remotely Sensed Ocean Color

Table 1

Cruise	Zone	Number of Stations	Optical Measurement
Vislab/SIO	Crater Lake Gulf Stream Gulf of Calif.	9	E_d, E_u[1]
SCOR/UNESCO Discoverer (1970)	Sargasso Sea Central–East Pacific	18	E_d, few E_u[2]
''	''	23	E_d, E_u[3]
Cineca–Charcot II (1971)	Upwelling off Mauritania	10	E_d, E_u[4]
Harmattan (1971)	Central–East Atlantic	16	E_d, E_u[4]
Cineca–Charcot V (1974)	Upwelling off Mauritania	30	E_d, E_u[5]
Guidom (1976)	South of Cape Verde Islands	11	E_d, few E_u[6]
Mares–Beagle II (1972)	Galapagos Islands	5	E_d[7]
Antiprod (1977)	South Indian and Antarctic	13	E_d, E_u[8]
4 Cruises (1974–76)	Japan Sea and Coastal Zones	20	E_d, E_u[9]
2 Cruises (1978)	Tokyo Bay Hisatsuru	13	E_d, E_u[10]

Interpretation of Remotely Sensed Ocean Color

Table 1 (Cont.)

Cruise	Zone	Number of Stations	Optical Measurement
Researcher (1977)	Gulf of Mexico	10	E_d, E_u, L_u^{11}
Crockett (1978)	Lake Erie	2	E_d, L_u^{11}
Gyre (1978)	Gulf of Mexico	10	E_d, E_u, L_u^{11}
New Horizon (1979)	Southern Calif. Bight	10	E_d, E_u, L_u^{11}
Oceanus (1979)	Western North Atlantic	12	E_d, E_u, L_u^{11}
David Starr Jordan (1979)	Eastern Pacific	13	E_d, L_u^{11}
Oceanographer (1980)	North Pacific (Japan–Seattle)	13	E_d, L_u^{11}
Desteiguer (1981)	San Diego to Hawaii	2	E_d, L_u^{11}
Fos (1979)	Fos Gulf, Etang Berre	8	E_d, E_u^{12}
Emicort (1977)	off Marseille Etang Berre	5	E_d, E_u^{12}
C-Fox (1979)	Vancouver Island	10	E_d, E_u^{12}

Table 1 (Cont.)

Cruise	Zone	Number of Stations	Optical Measurement
Pacific Clipper (1975)	Southern Calif. Bight	7	E_d, L_u^{13}
MTS (1977)	Chesapeake Bay	5	E_d, E_u, L_u^{13}
Gyre (1977)	Gulf of Mexico	10	E_d, E_u, L_u^{13}
Athena II (1978)	Gulf of Mexico	10	E_d, E_u, L_u^{13}
Velero IV (1979)	Gulf of Calif.	20	E_d, E_u, L_u^{13}
Athena II (1979)	Western North Atlantic	15	E_d, E_u, L_u^{13}
SAVS (1980)	Chesapeake Bay	2	E_d, E_u, L_u^{13}

1 Tyler and Smith (1970)
2 Smith (1973)
3 Morel (1973)
4 Morel and Caloumenos (1973)
5 Morel and Prieur (1976)
6 Morel and Prieur (1977)
7 Innamorati (1978)
8 Morel, Prieur, and Matsumoto (1978)
9 Okami, Kishino, and Sugihara (1978)
10 Okami et al. (1981)

---

11   Visibility Laboratory (Scripps Inst. of Oceanogr.)
     Unpublished
12   Laboratoire de Physique et Chimie Marines
     (Universite P. et M. Curie)
     Unpublished
13   National Earth Satellite Service (National
     Oceanic and Atmospheric Administration)
     Unpublished

Interpretation of Remotely Sensed Ocean Color

B. Atmospheric Effects

The retrieval of R or L_w from L_T requires an accurate estimate of the radiance scattered by the atmosphere and sea surface – L_1. The physics involved in such an estimate has been discussed in detail by Gordon (1978) (see also Gordon and Clark, 1980b, and Viollier et al., 1980). The main contributions to L_1 are scattering from the air (Rayleigh scattering) and scattering from particles suspended in the air (aerosol scattering). In principle, since the Rayleigh scattering can be determined from the atmospheric pressure at the surface, L_1 can be computed given the optical properties of the aerosol (e.g., see Quenzel and Kaestner, 1978). Unfortunately, in practice the aerosol properties are highly variable in both space and time and are in general unknown. It is necessary, then, to be able to estimate L_1 directly from the spacecraft observation.

The second term in Equation 1 can be rewritten to contain explicitly the water-leaving surface radiance $L_w(\theta,\emptyset)$ in Fig. 1. This is the radiance from the sun and sky which has been backscattered out of the ocean in a direction toward the sensor. The result, for a homogeneous atmosphere over an ocean for which R is independent of position, is

$$(9) \qquad RL_2(\theta,\emptyset)/(1 - rR) = tL_w(\theta,\emptyset),$$

with

$$(10) \qquad t = \exp[-(\tau_R/2 + \tau_{Oz} + (1 - \omega_A F)\tau_A)/\mu],$$

where $\mu = \cos\theta$, τ_R, τ_{Oz}, and τ_A are, respectively, the Rayleigh, Ozone, and aerosol optical thickness for the atmosphere, ω_A is the aerosol scattering albedo, and F is the probability that upon aerosol scattering a photon will be scattered through an angle less than 90 degrees. t is called the diffuse transmittance in contrast to the direct transmittance T, which is given by Eq. 10 with ω_A replaced by zero and the $\tau_R/2$ replaced by τ_R, and is hence less than t. (An alternate expression for t is provided in Tanre et al., 1979.) Physically, when a CZCS-type radiometer of infinite spatial resolution is aimed toward the pixel under examination the contribution to L_T from that pixel is TL_w; however, radiance from nearby pixels can be scattered toward the sensor, thereby increasing the apparent contribution from the pixel under examination to tL_w. As the spatial resolution of the sensor is decreased (i.e., the pixel size increased), some of the radiance from the pixel under examination can be scattered and still not leave the field of view of the sensor. In this case, the water-leaving radiance reaching the sensor is $t'L_w$, with $T < t' < t$. When there are real or apparent horizontal variations in R with very high contrast, such as between a beach and the ocean or due to the presence of clouds near the surface, it is possible for the contribution from adjacent pixels to occasionally far exceed that for the pixel under examination. These regions must be excluded from analysis.

The direct measurement of L_1 can be carried out only for those spectral regions for which R (or L_w) is essentially zero. Denoting a wavelength for which this is true by λ_o and the spectral band for which L_1 is desired by λ, it is necessary to be able to estimate $L_1(\lambda)$ from $L_1(\lambda_o)$. (Henceforth, the angles $\Theta, \emptyset$ will be suppressed in the arguments of all the radiance symbols. When an argument is given it refers to a specific wavelength; if no argument is given, then the equation or statement referencing the particular radiance is valid for all wavelengths.) This is facilitated by the observation that the Rayleigh and aerosol contributions to $L_1(\lambda)$ can be computed separately, i.e.,

(11) $L_1(\lambda) = L_R(\lambda) + L_A(\lambda)$,

where $L_R(\lambda)$ is the radiance for the same atmosphere and sea surface roughness in the absence of the aerosol, and $L_A(\lambda)$ is the radiance which would be observed in the absence of Rayleigh scattering. This result is strictly true in an atmosphere for which single scattering prevails; however, Gordon (1978) shows that it can be effectively employed even in a multiply scattering atmosphere as long as the Rayleigh optical thickness τ_R is less than about 0.25, the aerosol optical thickness τ_A is less than 0.6, and the radiometer is directed at least 20 degrees away from the specular image of the sun (the center of the sun glint pattern). The last requirement is satisfied by the CZCS under normal operating 'glint avoidance' procedures utilizing the sensor's tilting capability. The reason for the apparent lack of interaction between Rayleigh and aerosol scattering can be understood physically by noting the great difference in their scattering phase functions. (Note that this noninteraction allows the diffuse transmittance, t, to be written as in Equation 10.)

Since L_R can be computed from theory to any desired accuracy (however, see below, Section IV), Eq. 11 enables the determination of $L_A(\lambda_o)$ and

$$L_1(\lambda) = L_R(\lambda) + L_A(\lambda)$$

can be found if $L_A(\lambda)$ can be estimated from $L_A(\lambda_o)$. For small τ_A, $L_A(\lambda)$ is given approximately (single scattering) by

$$L_A \cong \omega_A \tau_A p_A(\mu,\mu_o) F_s T_{Oz}(\mu) T_{Oz}(\mu_o)/4\pi\mu$$

where

$$p_A(\mu,\mu_o) = P_A(\gamma_-) + [\rho(\mu) + \rho(\mu_o)]P_A(\gamma_+),$$

$$\cos\gamma_\pm = \pm\mu\mu_o + [(1-\mu^2)(1-\mu_o^2)]^{1/2}\cos(\phi-\phi_o),$$

$\mu = \cos\theta$, $\mu_o = \cos\theta_o$, ϕ_o is the azimuth of the sun, and ρ is the Fresnel reflectivity of the sea surface (air to water). $P_A(\gamma)$ is the phase function for aerosol scattering at the angle γ, F_s is the extraterrestrial solar irradiance, and the $T_{Oz}(\mu)$'s are the slant path transmittance through the Ozone, i.e., $T_{Oz}(\mu) = \exp(-\tau_{Oz}/\mu)$.

The ratio of the aerosol radiance at two wavelengths λ and λ_o is given by

$$S(\lambda,\lambda_o) \equiv L_A(\lambda)/L_A(\lambda_o)$$

(12)

$$= \frac{\omega_A(\lambda)\ \tau_A(\lambda)\ F_o(\lambda)\ p_A(\mu,\mu_o,\lambda)}{\omega_A(\lambda_o)\ \tau_A(\lambda_o)\ F_o(\lambda_o)\ p_A(\mu,\mu_o,\lambda_o)},$$

where $F_o(\lambda)$ is the extraterrestrial solar irradiance corrected for ozone absorption. Although this equation is not exact (because of multiple and interactive scattering) it is very useful in atmospheric correction. Consider a given aerosol type, defined here to be a given normalized particle size distribution and particle refractive index. Then, since the phase function $P_A(\gamma,\lambda)$ and the single scattering albedo $\omega_A(\lambda)$ are both independent of the total aerosol concentration and $\tau_A(\lambda)$ is directly proportional to the concentration, $S(\lambda,\lambda_o)$ will be independent of the concentration to the extent to which Eq. 12 is valid. Radiative transfer calculations suggest that the error in Eq. 12

due to multiple scattering rarely exceeds 10%.
Equation 12 shows that surface measurement of τ_A
can be used in atmospheric correction only if the
aerosol phase function and single scattering albedo
are both independent of wavelength. Since
$P_A(\gamma, \lambda)$ depends somewhat on λ, $S(\lambda, \lambda_0)$
will vary slightly over a CZCS scan line; however, it
seems reasonable to take as a working hypothesis that
$S(\lambda, \lambda_0)$ will be nearly constant over a CZCS
scene for which a single aerosol type prevails.

This hypothesis was first tested with CZCS
imagery by Gordon et al. (1980) by choosing λ_0 =
670 nm and calculating $S(\lambda, \lambda_0)$ directly from
ship measurements of L_u and determination of tL_w,
i.e.,

$$(13) \qquad S(\lambda, \lambda_0) \;=\; \frac{L_T(\lambda) \;-\; L_R(\lambda) \;-\; t(\lambda)\, L_w(\lambda)}{L_T(\lambda_0) \;-\; L_R(\lambda_0)}.$$

The same value of $S(\lambda, \lambda_0)$ was then used for the
entire image. The results suggest that the method is
viable and the approach has been implemented into the
NASA CZCS processing system. It should be noted that
the requirement of the existence of a spectral band
for which $L_w(\lambda_0) = 0$ places a restriction on the
waters for which this particular scheme is applicable.

III. IN – WATER ALGORITHMS

Before embarking on a detailed discussion of the
various in–water algorithms, a preliminary discussion
of the ocean constituents which can influence the
color of the water is useful. Considering the rich
assortment of absorbing and/or scattering agents which
are present in sea water (Figure 2), there is
virtually no hope that the concentration of all of the
individual constituents can be remotely estimated with
precision. In fact, it is not clear that even the
relative concentrations of the various algal color
groups can be estimated through accurate measurement
of $R(\lambda)$ at the sea surface. Experience has shown
that the particulate matter found in the open ocean in
large enough concentrations to produce observable
optical effects are principally living algal cells
(phytoplankton) and their associated detrital material
(mainly particulate, but also dissolved). These may
also be present in coastal and/or shallow areas along
with (inorganic) sediments resulting from land
drainage, (more or less mineralized) sediments from
the bottom which are resuspended by the action of
waves and tides, and organic (natural or
anthropogenic) sediments. Near the mouths of large
rivers such as the Amazon or Mississippi and in low
salinity seas (such as the Baltic, Canadian inlets,
etc.), dissolved organic material called yellow
substances (y.s.) can also strongly influence the
color of the water. In the light of their complexity,
realistic goals for ocean color remote sensing must be
developed. Thus far, the principal goals have been to
provide an indication of the concentration of algal
cells through an estimate of the concentration of
their absorbing pigments (phytoplankton pigments), and

to provide an estimate of the seston concentration. Seston is defined to be the total concentration of suspended particulate matter (SPM) including phytoplankton and their detrital material, if present. Seston concentration is determined gravimetrically and expressed as dry mass per unit volume, usually g/m^3. The term seston should not be confused with 'turbidity,' which has no precise meaning, but which we shall use restrictively here to describe the 'sediment load' in coastal/shallow water. Thus in what follows, the term seston refers to all suspended material, while turbidity refers to seston minus phytoplankton and their associated detrital material.

Morel and Gordon (1980) pointed out three different approaches for using measurements of spectral radiance (or irradiance) to estimate the concentrations of the various constituents of the water: (i) an empirical method, (ii) a semiempirical method, and (iii) an analytical method. In the empirical method the ratio (or difference) of upwelled radiance at two wavelengths is statistically related to the concentration of the particular constituent involved. The pigment and seston algorithms for satellite use described below are derived using this method. In the semiempirical method an attempt is made to separate the effects of the various constituents on the optical properties of the medium. This is accomplished by using statistical means to split the irradiance attenuation coefficient (K_d) into partial coefficients which describe the influence of water, chlorophyll a and covarying material, and material which does not covary with chlorophyll a. This method is used to derive the 'K − algorithm' described below. The analytical method directly uses the results of radiative transfer theory, and expresses the absorption and backscattering coefficients in terms of the constituents of the water. Then, given the values of the specific absorption and backscattering coefficients, and measurements of $R(\lambda)$ at several wavelengths, the system is inverted to provide C_i. Thus far this method has been applied only to measurements made at the sea surface (Morel, 1980) or from low flying aircraft (Jain and Miller, 1976).

(Actually, Jain and Miller used the 'two - flow'
method mentioned above rather than the rigorous
expressions from radiative transfer theory; hence it
would perhaps be more accurate to describe their
analysis as semiempirical.)

A. The Phytoplankton Pigment Algorithms

In the open ocean and most upwelling areas the
principal process influencing the optical properties
of the water is the absorption produced by pigments
contained in phytoplankton, and their immediate
derivatives. Thus it is to be expected, as proposed
by Clarke et al. (1970), that the chlorophyll a
concentration in surface waters could be estimated
from the 'color' of the water. This is biologically
useful (1) since the chlorophyll a concentration has
for many years been used as an indication of algal
bio-mass in productivity studies (see, e.g., Platt et
al., 1977), and (2) because it plays a central role in
the process of photosynthesis and, hence, primary
productivity of the water. Indeed, the rate of
photosynthesis is usually expressed in mg carbon per
cubic meter transferred to carbohydrate per unit
chlorophyll a concentration per unit time (Parsons and
Takahashi, 1973).
The concept of using ratios of radiances at
various wavelengths to estimate the chlorophyll
concentration in surface waters apparently originated
with Clarke et al. (1970). Ramsey and White (1973),
Clarke and Ewing (1974), Arvesen et al. (1975), and
Hovis and Leung (1977) also attempted to relate
radiance ratios to the chlorophyll concentration.
Usually these ratios involve wavelengths near the
maximum in the chlorophyll a absorption (440nm) and
the minimum in chlorophyll a absorption (560nm). It
was soon realized that due to the similarity between

the absorption spectrum of chlorophyll a in the blue
part of the spectrum, and one of its degradation
products, phaeophytin a, it would be impossible to
develop a means of estimating the concentration of
chlorophyll a alone with an instrument such as the
CZCS, which employs a small number of spectral bands.
Hence, relationships between the sum chlorophyll a
plus phaeophytin a (henceforth called phytoplankton
pigments and denoted by C) and the radiance exiting
the ocean were sought.

The first such algorithm was given (although not
as such) by Morel and Prieur (1977), who presented a
graph relating

$$\rho(440,560) \equiv R(440)/R(560)$$

to C. It was clear from their data that a rough
relationship

$$(14) \qquad C = A[\rho(440,560)]^B$$

could be established. In the same paper they
classified ocean water according to the relative
importance of phytoplankton and their covarying
detrital products compared to various inorganic and
organic sediments, 'Case 1' waters being those for
which phytoplankton and their derivative products play
a dominant role in determining the optical properties
of the ocean, and 'Case 2' waters those for which the
inorganic and/or organic sediments make an important
or dominant contribution to the optical properties.
This is summarized in Figure 2.

Waters ranging from oligotrophic (very low
pigment content) to eutrophic waters (very high
pigment content) belong to Case 1 provided that the
agents 4, 5, 6 and 7 do not exert a significant
influence; the always-associated agents 1, 2 and 3
determine the optical properties. According to
Bricaud, Morel and Prieur (1981), the measurable

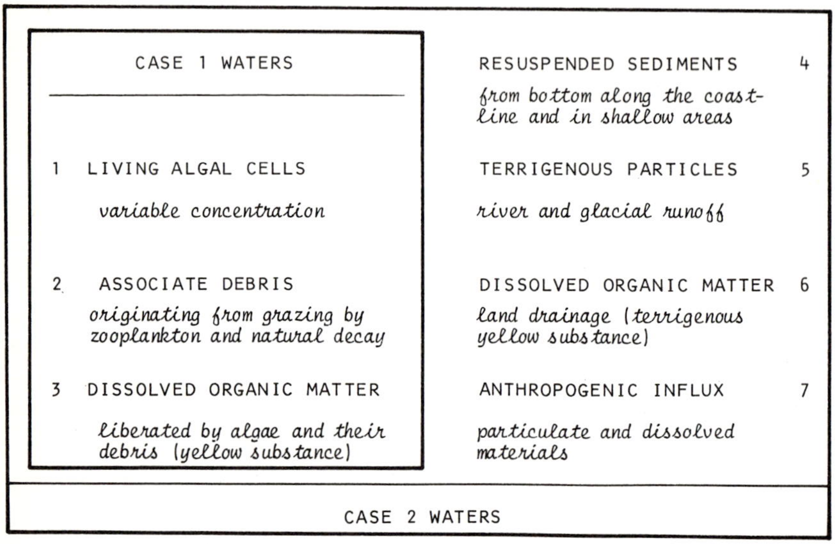

CASE 1 WATERS	RESUSPENDED SEDIMENTS 4
	from bottom along the coast-line and in shallow areas
1 LIVING ALGAL CELLS	TERRIGENOUS PARTICLES 5
variable concentration	*river and glacial runoff*
2 ASSOCIATE DEBRIS	DISSOLVED ORGANIC MATTER 6
originating from grazing by zooplankton and natural decay	*land drainage (terrigenous yellow substance)*
3 DISSOLVED ORGANIC MATTER	ANTHROPOGENIC INFLUX 7
liberated by algae and their debris (yellow substance)	*particulate and dissolved materials*

CASE 2 WATERS

Figure 2.   The main substances which determine the optical properties of a water body by absorption (dissolved material) or absorption and scattering (particulate matter). Case 1 waters are defined as such waters solely influenced by the components (always-associated) 1, 2 and 3. Waters containing at least one of the other components (among 4, 5, 6 and 7) are classified as Case 2 waters.

influence of 'marine' yellow substance, 3, (i.e., a
by-product of algae degradation) remains weak, even in
eutrophic areas.

Case 2 waters may (or not) contain the components
1, 2 and 3. Waters depart from Case 1 to enter into
Case 2 because of i) their high turbidity (sediment
load) due to the influence of 4 and/or 5 (they are
then sediment-dominated Case 2 waters); ii) their
high terrigenous yellow substance content (6) (they
are then yellow-substance-dominated Case 2 waters, or
gilvin dominated, according to Kirk, 1980); and iii)
their cumulated influence. Human activity, urban
sources, industrial wastes, (7), can also create
Case 2 waters, or superimpose their effects on
existing Case 2 waters.

Oceanic waters, as a rule, form the Case 1
waters. These waters, however, are also present even
in coastal areas in the absence of terrigenous influx
(arid climate) and of the continental shelf.
Eutrophic Case 1 waters occur in certain upwelling
regions, when the upwelled waters appear offshore,
over the outer shelf or shelf break. When they appear
over the inner shelf, they are often transformed into
Case 2 waters as the sediment resuspension, mainly
caused by waves and vertical mixing, maintains a high
turbidity. Both these situations are encountered, for
instance, along the N.E. African coast (see, e.g.,
Barton et al., 1977; Morel, 1982). Case 2 waters of
diverse kinds are normally encountered in coastal
zones (estuaries, shelf areas, inlets, etc.) and
possibly far from the coast in the case of extended
shelves or shallow banks.

Finally, note that among the constituents (1 to
7) considered, aeolian and meteoric dusts as well as
zooplankton have not been represented for the reason
that they have a negligible influence upon the optical
properties.

In contrast to Case 1 waters, sediment-dominated
Case 2 waters show relatively higher scattering,
which, in general, does not covary with phytoplankton.
At high phytoplankton concentrations, Case 1 waters
would appear dark green, while Case 2 waters would
appear a bright-milky green. Yellow

substance-dominated Case 2 waters, with relatively low
scattering, also appear dark, but more brown than
green.

Applying linear regression techniques on the
log-transformed ρ and C data Morel (1980) found that
A and B in Eq. 15 were, respectively, 1.62 mg/m^3 and
-1.40, and a coefficient of determination $r^2 = 0.76$.
When the Case 2 waters were excluded from the
analysis, it was found that A = 1.92 mg/m^3, B =
-1.80, and r^2 increased to 0.97 (See Table 2 and the
caption for Figure 3b).

Gordon and Clark (1980b) developed several
algorithms specifically for the CZCS spectral bands
using measurements carried out in very diverse waters
(examples from both Case 1 and Case 2 were
represented). These algorithms were of the form

$$(15) \qquad C = A[r_{ij}]^B,$$

with

$$r_{ij} = L_u(\lambda_i)/L_u(\lambda_j),$$

where the subsurface radiances, $L_u(\lambda)$, were
typically measured at one meter depth. In comparison
with Eq. 15, for λ_i = 440nm and λ_j = 550nm it
was found that A = 0.50 mg/m^3 and B = -1.27 with an
r^2 of 0.98. Even though ratios of radiance and
ratios of reflectances irradiance are not identical
(recall that Q is somewhat spectrally dependent), and
the wavelength used in ratios differ slightly, it is
comforting to note the similarity in the Gordon and
Clark and the Morel values for B. Since the launch of
the Nimbus-7, the data base which was used in the
Gordon and Clark study has been considerably expanded
to include the Sargasso Sea, the U.S. East Coast
including Georges Bank, the Gulf of California, the
West Coast of Baja as well as the original Gulf of
Mexico, Chesapeake Bay and Southern California Bight.

Clark (1981) has presented an analysis of this data with several improvements over the previous work. First, from L_u measurements at several depths, the upwelled radiance attenuation coefficient was determined, allowing propagation of the measurements at a depth of one meter to the surface. Next, the surface values of L_u were weighted by the spectral response of the CZCS bands. Finally, the regressions were carried out as above on the log–transformed equation

(16) $$C_f(\lambda) = A[r_{ij}]^B,$$

where C_f (called the optically weighted pigment concentration) is given by Eq. 7 with $\lambda = 520$nm, and λ_i and λ_j refer to the CZCS spectrally weighted radiances. It is significant to note that the analysis indicated no statistical difference between C_f at 520 nm and the surface pigment concentration. Also, only small differences were found between the A and B coefficients for the various (λ_i, λ_j) combinations above and those in Gordon and Clark (1980b). Comparison with Morel (1980) suggests that the Clark and the Gordon–Clark regression lines are strongly influenced by Case 2 waters. This can be confirmed by comparing Figure 3a, which shows Clark's data and regression line for the (443,550) wavelength combination along with Morel's regression for the (440,560) combination with Figure 3b, which shows Clark's regression line superimposed over Morel's data. (Note: in this comparison the difference between reflectance ratio and radiance ratio has been ignored). Clark's regression is clearly influenced by Case 2 waters (excluded from Morel's regression analysis), i.e., turbid waters influenced by varying amounts of sediment (Clark's high pigment concentrations were observed in Chesapeake Bay and the Mississippi Delta). Conversely, Clark's points from the Southern California Bight correspond well to Morel's Case 1. These algorithms are summarized in Table 2.

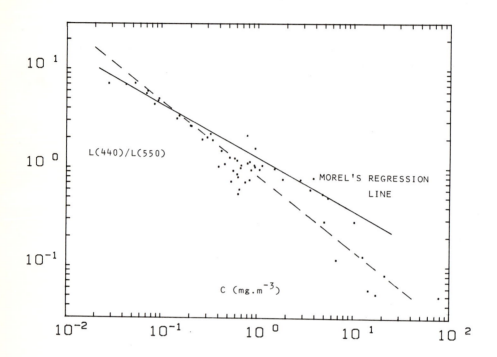

Figure 3a. Radiance ratio, $L_u(440)/L_u(560)$, plotted vs. pigment concentration (Clark's data, 1981) and regression line (dashed line). Morel's regression line comes from Fig. 3b (and represents the algorithm 3, Table 2).

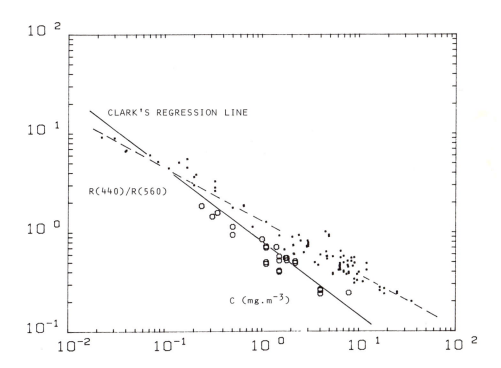

Figure 3b. Reflectance ratio, R(440)/R(560), plotted vs. pigment concentration. With respect to a previously published figure (Morel, 1980), 27 additional data have been included. These additional data were obtained around Vancouver island (C–Fox experiment), some of them in eutrophic waters. By using an exclusion test ($\rho > 0.91C^{-0.575}$) to select only Case 1 waters, and then performing the regression analysis, the relationship $C = 1.71\rho^{-1.82}$ ($r^2=0.954$ N=92) is obtained (dashed line), which does not significantly differ from algorithm 3, Table 2. Clark's regression linecomes from Fig. 3a (and algorithm 5, Table 2).

Table 2

Algorithms proposed by various investigators and having the form

$$M = Ar_{ij}^B \quad \text{with} \quad r_{ij} = L_u(\lambda_i)/L_u(\lambda_j)^*$$

Pigment algorithms: $M = C = (Chl\ a + Phaeo\ a)$ concentration (mg/m^3)

	λ_i	λ_j	A	B	r^2	Authors
(1)	443	550	0.50	-1.27	0.98	Gordon-Clark(1978-80)
(2)	440	560	1.62	-1.40	0.76	Morel (1978-80) (Case 1 + Case 2)
(3)	440	560	1.92	-1.80	0.97	Morel (Case 1 only)
(4)	443	550	0.78	-2.12	0.94	Smith-Wilson (1981)
(5)	443	550	0.77	-1.33	0.91	Clark (1981)
(6)	443	520	0.55	-1.81	0.87	''
(7)	520	550	1.69	-4.45	0.91	''
(8)	520	670	43.85	-1.37	0.87	''

Table 2 (Cont.)

'K' algorithms: $M = (K_{490} - 0.022)$ (9) (m^{-1})

or $M = (K_{520} - 0.044)$ (10)

	λ_i	λ_j	A	B	r^2	Authors
(9)	443	550	0.088	−1.491	0.90	Austin-Petzold (1981)
(10)	443	550	0.066	−1.398	0.995	''

Seston algorithms: $M = S$ = seston concentration (g/m^3)

(11)	440	550	0.398	−0.88	0.92	Clark, et. al., (1981)
(12)	440	520	0.331	−1.09	0.94	''
(13)	520	550	0.759	−4.38	0.77	''
(14)	443	550	0.24	−0.98	0.86	Clark-Baker (1981, per. comm.)
(15)	520	550	0.45	−3.30	0.86	''
(16)	520	670	5.30	−1.04	0.85	''

* or $r_{ij} = R(\lambda_i)/R(\lambda_j)$ for algorithms (2) and (3). When generating algorithms (9) and (10), data for $E_u(\lambda_i)/E_u(\lambda_j)$ and $L_u(\lambda_i)/L_u(\lambda_j)$ have been pooled.

35

The algorithms proposed thus depend on the set of data considered and on whether or not Case 2 waters are excluded. This is of minor importance, however, since there is no need for a 'universal' and ubiquitous algorithm. Different coefficients (A and B) can be used according to the region under consideration and the a priori knowledge of the situation; for instance, Smith and Wilson (1981) use a particular algorithm developed for the Southern California Bight (included in Table 2). Moreover, the iterative method of atmospheric correction, described below, allows $L_w(670)$ to be estimated. If high values are found indicating Case 2 waters, the computation can be switched to an algorithm for turbid coastal waters or to a site-specific algorithm. In any event, a better tightness of fit and a universal and more accurate algorithm is obviously obtained for oceanic Case 1 waters. The standard error of estimate of the regression suggests that the pigment concentration can be estimated to within a factor of two ($\pm 40\%$) for these waters over the full range of pigment concentration ($0.02 - 20$ mg/m^3), and is considerably reduced if only part of the range is considered, e.g., $0.02 - 0.7$mg/m^3. The uncertainty is approximately doubled if certain Case 2 waters are considered together with Case 1 waters, as results from Clark's (1981) study for U.S. waters. However, it remains that coastal waters diversely influenced by land run off, human activity, etc., are more unpredictable.

B. The 'K' Algorithms

In the absence of vertical stratification K_d, the attenuation coefficient of downwelling irradiance, is nearly independent of depth. As mentioned above, an important property of this quantity in remote sensing lies in its relationship to the penetration depth z_{90}. In a homogeneous ocean only 10 percent of the photons which are backscattered out of the ocean reach depths greater than $1/K_d$ (Gordon and McCluney, 1975); thus measurement of K_d provides an estimate of the extent to which the remote sensor can 'see' into the ocean.

The irradiance attenuation coefficient is also of considerable importance in primary productivity studies. Since photosynthesis is a quantum process it is important to convert the radiant energy available for photosynthesis to quanta through

$$E_q(\lambda) = \lambda E_d(\lambda)/hc.$$

where c is the speed of light and h is Planck's constant. The total radiant energy available for photosynthesis (see, e.g., Morel, 1978) is then

$$PAR(z) = \int_{350}^{700} E_q(\lambda)d\lambda$$

and the attenuation coefficient for downwelling quanta is defined by

$$K_q = -d(Ln[PAR(z)])/dz.$$

Normally photosynthesis takes place significantly above the depth, $z_q(1\%)$, at which 1% of the number of quanta incident at the surface remain. The region between the surface and this depth is called the euphotic zone, and a remote estimate of its depth is of obvious importance, and can be effected under the assumption of homogeneity within this zone and by using relationships between K_q and $K_d(\lambda)$.

Smith and Baker (1978a, 1978b) were the first to apply the semiempirical method mentioned above in an attempt to split K_d into components and relate its

spectrum to the pigment concentration in the water.
(K_d, an apparent property, is, unlike a, not
rigorously summable over the constituents, thus the
partition is an approximation.) They limited their
study to Case 1 waters, and partitioned $K_d(\lambda)$
according to

$$K_d(\lambda) = K_w(\lambda) + k_1(\lambda)C$$

for $C < 1\text{mg/m}^3$, and

$$K_d(\lambda) = K_w(\lambda) + K'(\lambda) + k_2(\lambda)C$$

for $C > 1\text{mg/m}^3$, where C (C_k in Smith and Baker's
notation) is the pigment concentration averaged over
the penetration depth, and K' is the attenuation not
attributable directly to phytoplankton or their
covarying detrital material (K' is taken to be
independent of C). The necessity of using two separate
equations to relate C and K_d is a manifestation of
the fact that, although phytoplankton and their
detrital material covary, they do not covary linearly.
At low pigment concentrations [$C < 1 \text{ mg/m}^3$] there is
relatively more covarying detrital material than
viable phytoplankton and so, in this regime, both
influence $K_d(\lambda)$ making $k_1(\lambda)$ difficult to
interpret. On the other hand, at high concentrations
[$C > 1 \text{ mg/m}^3$] the concentration of covarying
detrital material is relatively lower. $k_2(\lambda)$ could
be interpreted as the 'specific diffuse attenuation
coefficient' of phytoplankton pigments. $k_2(\lambda)$ is,
on the average, four to five times smaller than the
corresponding $k_1(\lambda)$ value. (Note that the
division between these two regions is not sharp, and
$C = 1 \text{ mg/m}^3$ is simply an estimate; its value, of
course, determines the value of $K'(\lambda)$.) This can be
related to the true specific absorption coefficient of
phytoplankton $a_{ph}(\lambda)$ by noting that the
downwelling irradiance attenuation coefficient is

Interpretation of Remotely Sensed Ocean Color

approximately given by

$$K_d \cong K \quad \text{with} \quad K = -d(LnE)/dz = a/\bar{\mu}.$$

Thus, at high concentrations that part of a which co-varies with C must be $a_{ph}(\lambda)$, so,

$$k_2(\lambda) \cong a_{ph}(\lambda)/\bar{\mu}.$$

Comparison between k_2 and the specific absorption coefficient of phytoplankton determined by Morel and Prieur (1977), also for waters with high pigment concentrations, shows that this is indeed the case, and that the factor $1/\bar{\mu}$ appears to be of the order of 1.3 ± 0.15 in the $440 - 620$ nm spectral region. Between 640 and 680 nm, in the vicinity of the red absorption peak, where the k_2 values are less reliable, k_2 is lower than, and hence disagrees with, a_{ph}.

The first indication that irradiance and quantal attenuation coefficients could be estimated from remote water color measurements is found in Jerlov (1974). In that paper a relationship was observed between the 'color ratio' (the ratio $F = L_u(447)/L_u(519)$ measured at 1 m depth) and the depth at which 30% of surface downwelling irradiance at 465 nm remain, and also the depth at which 10% of surface downwelling quanta remain. Højerslev and Jerlov (1977) (see also Højerslev, 1980) presented a simple relationship between $z_q(1\%)$ and F,

$$z_q(1\%) = 9.8 + 25.2F,$$

or conversely,

$$F = 0.040 z_q(1\%) - 0.389.$$

Interpretation of Remotely Sensed Ocean Color

This was revised by Højerslev (1980) to read

$$F = 0.00013 z_q(1\%)^2 + 0.017 z_q(1\%) + 0.14.$$

Similar relationships, developed for the CZCS bands, should be applicable to the remote estimate of $z_q(1\%)$, at least when it can be expected that the ocean is homogeneous within the euphotic layer.

Austin and Petzold (1981) have developed an algorithm which relates the attenuation coefficients for upwelling radiance and/or irradiance to the ratio of upwelled radiances at two wavelengths. The upwelling attenuation coefficients are much easier to measure with precision than the associated downwelling coefficients and much less affected by the radiance distribution incident on the sea surface and on the wave-induced surface roughness. (To the extent that the upwelled subsurface radiance distribution is diffuse, these upwelled attenuation coefficients are 'diffuse' attenuation coefficients, i.e., they would describe the attenuation of diffuse light propagating through the medium.)

From a statistical analysis of 88 selected upwelled spectral radiance and irradiance measurements, Austin and Petzold (see Table 2) find relationships of the form (the subscript u, in K_u, is omitted)

$$K(\lambda) - K_w(\lambda) = A(r_{ij})^B$$

for $\lambda = 490$ and 520 nm, where as before, r_{ij} is $L(\lambda_i)/L(\lambda_j)$ or $E_u(\lambda_i)/E_u(\lambda_j)$. From the K values, the K_w values for pure water at the same wavelength are subtracted. These K_w values are estimated from the approximate relationship (the approximation includes the assumption that $\bar{\mu}$ is close to 1 and leads to a lower limit for K_w),

$$K_w = (a)_w + (b_b)_w,$$

with $(a)_w$ and $(b_b)_w$ from Morel and Prieur (1977)
and Morel (1974), respectively. The high coefficients
of determination for the above regressions allow, in
most cases, K (at 490 and 520 nm) to be determined to
within about 25% over the range from K_w to about
0.3 m^{-1}, corresponding to pigment concentrations
from zero to about 2 mg/m^3.

In the process of developing these algorithms,
Austin and Petzold noticed that excellent correlations
existed between ratios of subsurface radiances at the
various CZCS spectral bands. In particular, they found
a good relationship between $L_u(443)/L_u(670)$ and
$L_u(443)/L_u(550)$, (see below, Section E), which has
been used by Smith and Wilson (1981) in their
iterative approach to atmospheric correction (see
below, Section IV).

C. The Seston Algorithms

Basically, algorithms for seston were obtained
through the empirical (statistical) method. The ratios
of spectral radiances that are measured near the
centers of the CZCS bands were studied, along with the
seston concentration through a regression analysis on
the log-transformed data. The resulting algorithms
were obviously of the same form as the phytoplankton
algorithms, i.e.,

$$S = A(r_{ij})^B.$$

The coefficients A and B, as obtained by Clark, Baker,
and Strong (1980), are given in Table 2. These

coefficients are slightly modified (Clark and Baker, personal communication) if the radiances are weighted by the normalized spectral responses of the CZCS bands.

The similarity between the seston and pigment algorithms implicitly expresses that a tight correlation must exist between both of these quantities. As a matter of fact, such correlations have been observed by directly comparing the seston and pigment concentrations measured in near-surface waters (see below, Section E).

In Case 1 waters, seston originates from photosynthetic and biological activity; thus it appears reasonable to expect an approximate covariation of the total seston with the algal biomass (described by the pigment concentration). When the supply of nonbiogenous material is added to and often exceeds the biogenic production, the C-S correlation vanishes and the seston algorithm on which it rests, fails. In such situations (turbid Case 2 waters) the spectral behavior of $L_w(\lambda)$ (and hence the radiance ratios) is less significant than the high reflectance which characterizes these waters. This reflectance is often sufficiently high to be detected even with the insensitive MSS on LANDSAT-1 (e.g., see Maul and Gordon, 1975). The high reflectance everywhere within the spectrum could be used to indicate and to map turbid areas (assuming that the atmospheric correction can be effected over such highly reflecting waters).

D. The Analytic Algorithm

As mentioned above, the analytic algorithm consists of relating constituent concentrations to the optical properties of the medium, and hence to the reflectance of the ocean, using expressions derived from radiative transfer theory. Specifically, a and

b_b are related to the constituents through Equations 3 - 5, and Eq. 6 is used to relate these to $R(\lambda)$. Several equations can be written (corresponding to different wavelengths) and by inverting this system the algorithms become explicit (Morel, 1980). They theoretically allow the computation of the desired concentrations from the set of $R(\lambda)$ values. This multi-spectral analysis is necessary since there is no spectral band (in the visible) in which the influence of a single component can be isolated so its concentration could be inferred separately. The insufficient knowledge of the specific spectral absorption and backscattering values to be used in the equations limits, at present, the accuracy of such a method. Further improvements of algorithms, however, and attempts to retrieve independently several components from color measurements necessarily pass through an analysis of this kind.

To solve (at least partly) this problem, a study of the optical properties of the algal cells has been undertaken (Morel and Bricaud, 1981a). From computations using Mie theory, and also as evidenced by experience, it is clear that the backscattering of phytoplankton is very weak and strongly influenced by absorption. Also, the computations reveal that the concept of 'specific' absorption (absorption per unit concentration of pigment) is more complex than generally thought since it depends on the mean size of the cells (decreasing for increasing cell size), as well as on the intracellular pigment concentration (decreasing for increasing concentration). These results (Morel and Bricaud, 1981b) are important in ocean color modeling; for example, they enable a very simple model of the pigment algorithm for Case 1 waters to be constructed.

Consider hypothetical waters which would contain only phytoplankton. Since the backscattering of the phytoplankton is very small at low or moderate concentrations, the backscattering of the medium will be due to the water itself and, thus, independent of the phytoplankton concentration. In this case, from Eq. 6 the ratio of reflectances at two wavelengths becomes

$$\frac{R(\lambda_2)}{R(\lambda_1)} = \frac{b_{bw}(\lambda_2)[a_w(\lambda_1) + a_{ph}(\lambda_1)C]}{b_{bw}(\lambda_1)[a_w(\lambda_2) + a_{ph}(\lambda_2)C]}.$$

Using the spectral absorption coefficients determined for pure water ($a_w(\lambda)$) and the specific absorption for phytoplankton ($a_{ph}(\lambda)$) by Morel and Prieur (1977), the reflectance ratio for any two wavelengths can be derived as a function of the phytoplankton concentration (C). Figure 4 shows the results of such a calculation for 440 and 560 nm along with the Morel Case 1 algorithm. It appears that this very simplified model provides a realistic limit to the coefficient B (Eq. 14) for the pure Case 1 waters. Also, the overall upward shift of the theoretical curve relative to the experimental measurements suggests that the backscattering cannot be solely due to water. In fact if it is assumed that $b_b(\lambda_2) = b_b(\lambda_1)$, as a result of the presence of detrital particles, the theoretical computations and the Morel Case 1 algorithm are in reasonable agreement, i.e., they differ by less than a factor of 2 over the range $0.1 < C < 20$ mg/m^3. An improved model will be developed later (see below, Section E).

Several authors (Viollier et al., 1978; Viollier et al., 1980; Tassan, 1981) have proposed the use of radiance difference (rather than ratio) algorithms to estimate the constituent concentrations. The rationale is that by taking differences of radiances (after removal of the contribution from Rayleigh scattering) the severe requirements on the estimate of the aerosol contribution to the atmospheric effects can be relaxed. Also, partial compensation for other image-degrading effects such as sun glint and white cap reflection could be achieved. The simple model above can be used to demonstrate the dangers associated with the use of such an algorithm. Assume that the water contains only two components: phytoplankton and nonabsorbing inorganic suspended material with, respectively, concentrations C and s_i. Then in a typical coastal situation for which the backscattering of the phytoplankton and water

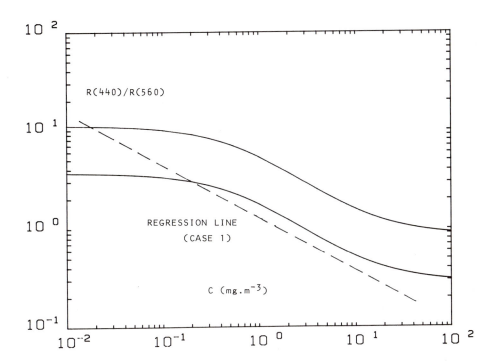

Figure 4. Variation of the reflectance ratio R(440)/R(560) with changing pigment concentration according to the predictions of a simplified model (see text).

would be negligible compared to that of the inorganic material,

$$R(\lambda) = 0.33b_b^s(\lambda)s_i/(a_w(\lambda) + a_{ph}(\lambda)C),$$

where $b_b^s(\lambda)$ is the specific backscattering coefficient of the inorganic material. This would be a good approximation to some Case 2 waters. The ratio $R(\lambda_2)/R(\lambda_1)$ is independent of s_i, to the extent that b_b^s is only slightly variable with λ, while the difference $R(\lambda_2) - R(\lambda_1)$ given by

$$0.33[R(\lambda_2)/R(\lambda_1) - 1] \ [b_b^s(\lambda_1)/(a_w(\lambda_1) + a_{ph}(\lambda_1)C)]s_i$$

is directly proportional to s_i. Thus for waters with the same pigment concentration but different quantities of noncovarying inorganic suspended material the reflectance ratios would be approximately the same, but the reflectance differences would not.

With respect to this model, actual situations are more complex. Inorganic suspended matter cannot be regarded as nonabsorbing, so an additional term, $a_s(\lambda)s_i$, has to be introduced into the above expressions, and also, b_b^s is not wavelength independent. Consequently, reflectance ratios are also sensitive to s_i (if not, algorithms for Case 1 waters could fully apply in Case 2 waters) and reflectance differences are not simply proportional to s_i. The aforesaid conclusion, however, remains valid because the reflectance differences are too much influenced by the change in backscattering (hence in s_i) to constitute a reliable index of a change in absorption.

E. Relationships Between the Algorithms

The algorithms derived from a statistical analysis of the log-transformed data (empirical method) are necessarily of the form:

(17) $M_k = A_k(r_{ij})^{B_k}$,

where r_{ij}, as previously, is the ratio of upwelling radiances (or irradiances) measured at two wavelengths λ_i and λ_j, and M is a marine parameter such as C, the pigment concentration, S, the seston concentration, or $[K(\lambda)-K_w(\lambda)]$, the attenuation coefficient for upwelling irradiance (from which the effect of water is subtracted). A_k and B_k are the constants which result from the regression analysis. The separate presentation of algorithms for various parameters, as given above, hides a pitfall. These algorithms, when obtained through the empirical method, are inevitably redundant expressions, all deriving from a unique independent relationship:

i) if several relationships as (17) have been produced in reference to a given marine property M by using several r_{ij} ratios, after eliminating M in the system of equations formed, relationships between the r_{ij}'s will be obtained;

ii)   conversely, given a ratio r_{ij} and a system of equations written with several algorithms which make use of this ratio to express several properties M, these will appear linked two-by-two, after eliminating r_{ij}. This simple reasoning also demonstrate that Case 2 waters, which allow seston and phytoplankton concentration to vary independently, cannot enter into this

scheme, characterized by a complete
determinacy.

The above discussion implies that the algorithms have
been considered as strict mathematical equations.
They are actually statistical products affected by a
certain degree of uncertainty ($r^2 < 1$) and, on the
other hand, they have been generated from different
data sets. For both these reasons, the algorithms
could form a system of incompatible equations which
cannot be inverted, or which would lead to meaningless
results. This question is examined in detail below.

1. Relationships Between the Radiance Ratios r_{ij}

By following the remark i) and considering the
system of algorithms (5) (6) (7) (8) in Table 2,
proposed by Clark (1981) to retrieve C from diverse
r_{ij} ratios, the following relationships appear when
C is eliminated:

$$L_W(443)/L_W(520) = 0.833[L_W(443)/L_W(550)]^{0.735}$$

$$L_W(550)/L_W(520) = 1.060[L_W(443)/L_W(550)]^{-0.299}$$

$$L_W(670)/L_W(520) = 0.052[L_W(443)/L_W(550)]^{-0.969}.$$

A direct study of the relationships existing
between the r_{ij}'s has been performed by Austin and
Petzold (1981) from a different and much larger body
of spectral upwelling radiances and irradiances. The
factors, A, and exponents, B, resulting from their
regression analysis and for the same ratios as above
are respectively:

$$A = 0.900 \qquad\qquad B = 0.742$$

$$A = 0.900 \qquad\qquad B = -0.258$$

$$A = 0.075 \qquad\qquad B = -0.919.$$

It is comforting to observe that these two sets of A
and B values are very consistent.

2. Seston–Pigment Relationships

According to the second remark and by coupling
algorithms for S and C, which are expressed by means
of the same r_{ij}, relationships between these
quantities can be established. Thus by coupling,
respectively, (14) with (5), (15) with (7), and (16)
with (8) (from Table 2,) the following expressions are
obtained:

$$S = 0.286 \; C^{0.705}$$

$$S = 0.304 \; C^{0.742}$$

$$S = 0.300 \; C^{0.759},$$

which again demonstrated an internal consistency
between the algorithms considered. Another test
consists of examining whether these relationships are
realistic by comparison with others obtained through a
direct correlation between simultaneous measurements
of S and C concentrations. For data from the Gulf of
Mexico, the Eastern Tropical Pacific, offshore and
coastal waters of Southern Alaska, and the New York
Bight area, Clark et al. (1980) found that over three
orders of magnitude in both S and C, these variables
were correlated ($r^2 = 0.79$) according to

$$S = 0.5 \ C^{0.75}.$$

Wide departures from this relationship (high and variable values of S with low C values) were found in the Lower Cook inlet (Alaska) due to a large amount of inorganic particulates introduced by glacial meltwaters (see Figure 4 in Clark et al., 1980). These data, not included in the regression, would be classified as turbid Case 2 waters. Some other data, less markedly anomalous, but apparently belonging to Case 2 waters (S $>$ 1 g/m^3 for C = 0.7–1.0 mg/m^3), were kept in the regression and are at the origin of the rather high value, 0.5, in the above expression.

A comparison with another source of data is possible since the scattering coefficient b, at 550 nm, has been studied along with C and is presented in Figures 5a and 5b. A somewhat arbitrary limit between Case 1 and Case 2 waters has been proposed by Morel (1980). After readjustment, this limit corresponds to (see legend Fig. 5a)

$$b = 0.45 \ C^{0.62} \qquad (b \ in \ m^{-1}).$$

Within the Case 1 domain, and over three orders of magnitude in C, the adopted relationship (valid for Case 1 waters only, Figure 5a) is

$$b = 0.30 \ C^{0.62} \qquad (b \ in \ m^{-1}).$$

According to Figure 6, an excellent linear relationship exists between b and S for near-surface waters and over two orders of magnitude. It is simply expressed by

$$b(m^{-1}) = S(g/m^3).$$

All the data obey this simple law within a factor of two. Thus b can be replaced by S in the above expressions; therefore:

(18) $S = 0.30 \, C^{0.62}$,

which compares favorably with the formulae derived from algorithms. The results displayed on Figure 5b (L. Prieur, personal communication) provide further insight into the C-S relationship in Case 1 waters (off Villefranche-sur-Mer, during the bloom which occurs in spring along the frontal-structures). The general trend in the distribution of the points suggests that a linear correlation must exist between the quantities b and C. Locally, and in a more restricted range of variation in C (from 0.5 to 5 mg/m^3 in this experiment) linear relationships are likely to be expected (see also Kiefer and Austin, 1974). This is not in contradiction with power laws and exponents around 0.6 and 0.7, which appear when a more extended range is considered, and which are accounted for by other observations. In Case 1 waters, the seston concentration (or b) is due not only to living algal cells but also to their detrital retinue. It has been recognized (see e.g., Hobson et al., 1973; Banse, 1977) that in waters with large standing crops of phytoplankton, the detritus is comparatively less abundant. The relative change in the proportions of dead and living particulates is at the origin of the nonlinear biological effect (Smith and Baker, 1978a, 1978b) and also could explain, in the case of C-S relationships, the presence of an exponent less than 1. The same explanation will be invoked later about the K-C relationships. For waters which do not follow the C-S correlation and belong to Case 2 for their high turbidity, the seston algorithms in Table 2 must fail. The spectral behavior of $L_u(\lambda)$, and hence radiance ratios, are less significant than the high reflectance which characterizes this kind of waters. Such a high reflectance, everywhere in the spectrum, could be used

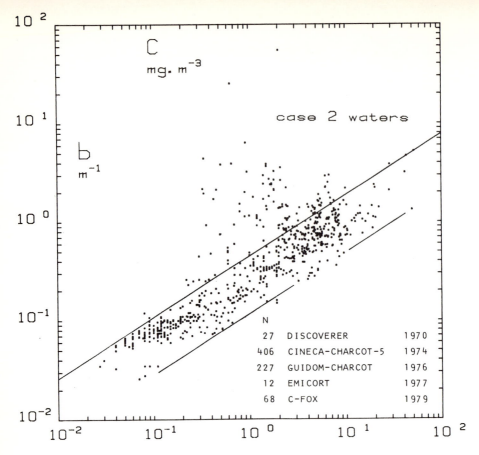

Figure 5a. Scattering coefficient b , at $\lambda = 550$ nm, plotted vs. pigment concentration C. Data from Guidom cruise (143), Emicort experiment (12) and C-Fox cruise (79) have been added to those previously presented (Morel, 1980). The inclusion of Case 1 waters (Guidom) with C varying from 0.05 to 1 mg/m^3 has led to move slightly upward that line which could separate the Case 1 from Case 2 waters. This line, which splits the figure into two panels, corresponds to $b = 0.45\ C^{0.62}$ (instead of $0.42\ C^{0.63}$ earlier). The band (lower panel) inside which Case 1 waters seem confined contains 506 points and regression analysis gives:

$$b = 0.246C^{0.59} \quad \text{or} \quad b = 0.258C^{0.65}$$

with $r^2 = 0.898$. Surface waters, in most cases, are represented by points close to the above upper limit. So an expression, $b = 0.3C^{0.62}$, can be considered as more representative for surface waters than the statistical product.

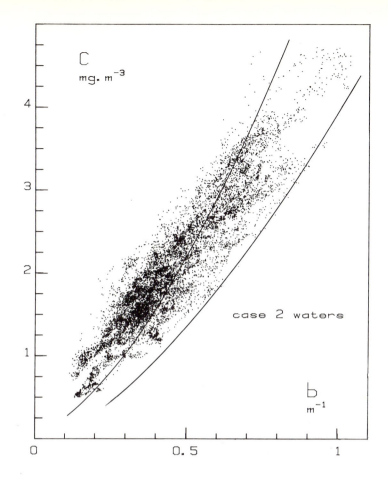

Figure 5b.   Plot resulting from continuous
measurements, numerically sampled each 10 seconds, of
b, at 620 nm (towed instrument) and of in-vivo
chlorophyll fluorescence.   About 16000 points are
plotted corresponding to 45 hours of records obtained
between March 24 and 28, 1980, along a zig-zag track
crossing blooms associated with a hydrologic front off
Villefranche-sur-Mer (~ 25 km).   The fluorescence
signal is converted into the chlorophyll a
concentration by using appropriate calibration factors
established each 30 minutes (sampling and conventional
spectrophotometric titration) (L. Prieur, unpublished
results, Prolig cruise).   The curve which envelops the
dots represents the limit between Case 1 and Case 2
waters ($b = 0.45C^{0.62}$), and the curve crossing the
cloud corresponds to $b = 0.3C^{0.62}$ (see legend, Fig.
5a).

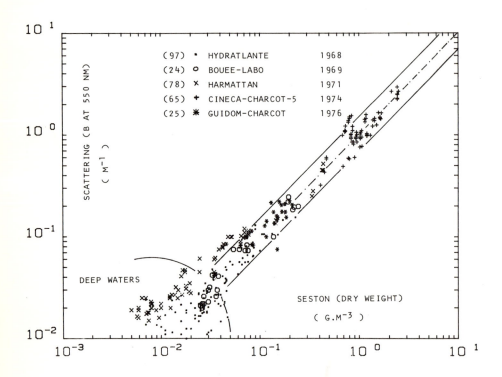

Figure 6. Scattering coefficient b , at λ = 550 nm, plotted vs. seston concentration (obtained by filtration using pre-weighted filters, rinsing, drying at 60°C and weighting). These data have been separately presented earlier (Morel, 1970 and 1973b; Morel and Prieur, 1976) except those obtained during the Guidom-Charcot cruise (Guinea Dome area).

to indicate and to map turbidity (assuming that
atmospheric correction can be effected over such
bright waters).

3. Pigment-'K' Relationships

In a way similar to that above, r_{ij} can be
eliminated between algorithms giving $K(\lambda)-K_w(\lambda)$
and C. By considering the algorithms (5) and (9) and
then (5) and (10) the following relationships are
found:

$$K(490) - 0.022 = 0.118 \, C^{1.121};$$

$$K(520) - 0.044 = 0.0878C^{1.052}.$$

Exponents greater than 1 imply that the specific
effect of phytoplankton upon attenuation increases
with increasing concentration. This result is not
satisfactory since the converse has been observed
through direct studies of K-C (Smith and Baker, 1978a
and 1978b; Morel, 1979) or of absorption-C relations
(Prieur and Sathyendranath, 1981). The nonlinear
biological effect upon the optical properties (for
Case 1 waters), as a result of a relative diminishing
role played by the associated detrital material when C
increases (see above), corresponds to an exponent less
than 1. The discrepancy between this ascertained fact
and the preceding expressions originates from the fact
that the data used by Austin and Petzold to generate
the K's algorithms belong mostly to Case 1 waters but
include a few Case 2 waters, whereas Clark's
algorithms mix, more or less equally, Case 1 with
Case 2 waters. If Morel's algorithm (3), restricted
to Case 1 waters, is now associated with (9) and (10),
differing relationships are obtained:

$$K(490) - 0.022 = 0.096 \ C^{0.828};$$

$$K(520) - 0.044 = 0.072 \ C^{0.777}$$

(for K(490) varying from 0.03 to $0.4 \ m^{-1}$, and K(520) from 0.05 to $0.4 \ m^{-1}$). The exponents less than 1 appear more realistic according to the above discussion, and agree with the results of a direct study of the correlations between $K-K_w$ and C (Morel, unpublished). In that study $K_d(\lambda)$ values have been computed for the whole euphotic layer (99 stations), i.e., between surface and z(1%), or surface and a depth smaller than z(1%) (51 stations), and studied along with the average pigment concentration (C) computed for the layer under consideration. The regression analysis, performed after having excluded the (28) stations unambiguously identified as Case 2 waters, leads to

$$K(490) - 0.0217 = 0.069 \ C^{0.702} \qquad (r^2=0.95 \text{ and } N=121);$$

(19)

$$K(520) - 0.0489 = 0.050 \ C^{0.681} \qquad (r^2=0.94 \text{ and } N=120).$$

K varies between values near pure water to $K(490) \cong 0.9 \ m^{-1}$ and $K(520) \cong 0.7 m^{-1}$, whereas C varies from 0.02 to about 40 mg/m^3 (see Figures 7a and 7b). It must be pointed out that Austin and Petzold have chosen to consider $K_u(\lambda)$, the attenuation coefficient for L_u or E_u, mainly because these values are less noisy by the light field fluctuations and less affected by the sun altitude than $K_d(\lambda)$, particularly when K_d is computed for the surface layer and over a small depth increment (from 0 to $z \cong 1/K_d$). In the above mentioned study, these drawbacks are avoided as K_d is computed from measurements just above the surface, $E_d(0^+)$, and at a greater depth ($z \cong 2$ to $5/K_d$). This procedure also allows the error due to inaccurate

determinations of the depth interval to be minimized.
Theoretically K_d and K_u differ according to

$$K_d - K_u = (1/R)(dR/dz),$$

but practically, and within the confidence interval in
the preceding regressions and algorithms, they are
directly comparable. Hence, the agreement between the
straightforward Morel results and those derived from
the Austin-Petzold data combined with algorithm (3) is
meaningful.

In the above K-C relationships, once again an
exponent around 0.7 appears. That value apparently is
characteristic of and adequately reproduces the
nonlinear biological effect in Case 1 waters in the
middle range of variations in C and over more than two
orders of magnitude.

Another way of ascertaining these exponents, and
simultaneously of verifying the consistency within the
set of relationships above, consists in using the
selected K-C and C-S relationships in order to predict
the behavior of $\rho(440,560) = R(440)/R(560)$ with
varying C. The result can thereafter be compared to
the statistical product, i.e. the algorithm (3) in
Table 2. The ratio of reflectances

$$\rho(440,560) = [b_b(440)/b_b(560)][a(560)/a(440)]$$

can be expressed with respect to $K_d(\lambda)$ by noting
that $a(\lambda)$ differs from $K_d(\lambda)$ approximately by
the factor $\bar{\mu}$ (see above) and that $\bar{\mu}$, which is
weakly wavelength dependent, disappears when forming
the ratio. Thus a reasonably good approximation for
ρ is

$$\rho(440,560) = [b_b(440)/b_b(560)][K_d(560)/K_d(440)].$$

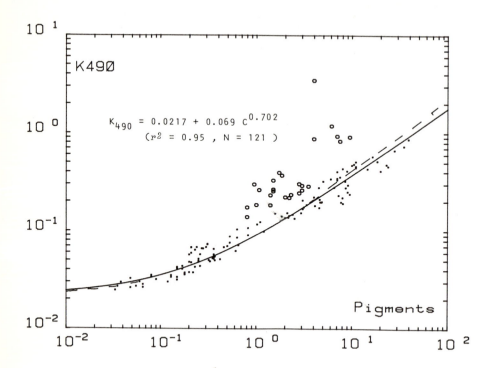

Figure 7a. The correlation analysis was performed between the log-transformed quantities C and K − K_w. The regression lines and the data are plotted in the plane K − C at a wavelength of 490 nm. The circles stand for Case 2 waters, excluded from regression. The results used in this study originate from the following experiments: Discoverer (29 data points with K computed for approximately the euphotic layer, i.e., the layer of thickness z=4.6/K, plus 14 data where K is computed for the surface layer, z < 2/K); Cineca II(7); Cineca V (18 + 14); Guidom (9); Antiprod (15) ; Fos (4) , C-Fox (17 + 23). When the difference K−K_w is not significant, the corresponding value is discarded. Separate regressions performed on the data concerning the euphotic layer or the near surface layer are not significantly distinct. The pooled data yield to a better fit and higher r^2 values.

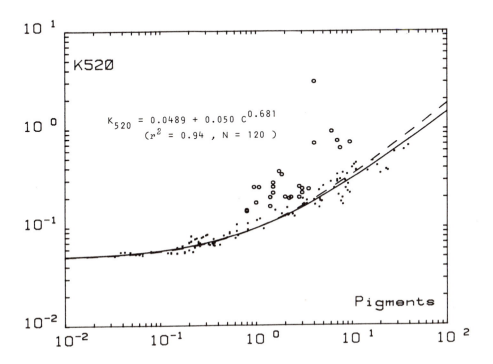

Figure 7b. **Same as Figure 7a at a wavelength of 520 nm.**

Statistical relationships for λ = 440 and 560 nm, similar to Eq. 19 have been obtained:

$$K_d(440) = 0.0168 + 0.1031C^{0.707} \quad (r^2=0.937,\ N=122);$$

(19')

$$K_d(560) = 0.0717 + 0.0390C^{0.640} \quad (r^2=0.947,\ N=101).$$

Relating b_b to C using Equation 18, ρ becomes explicitly

$$\rho = \frac{(b_w/2+\overline{b}_b 0.3C^{0.62})_{440}(0.0717+0.039C^{0.64})}{(b_w/2+\overline{b}_b 0.3C^{0.62})_{560}(0.0168+0.103C^{0.71})},$$

with b_w at 440 and 560 nm, respectively, equal to 4.95 and $1.80 \times 10^3 \text{m}^{-1}$ (Morel, 1974). After rearranging, the expression becomes

$$\rho = \frac{17.95+21.50C^{0.62}+9.75C^{0.64}+11.7C^{1.26}}{1.51+5.04C^{0.62}+9.27C^{0.71}+30.9C^{1.33}},$$

if $\overline{b}_b$ is assumed to be constant and equal to 10^{-2} at both wavelengths. This relationship is graphically shown on Figure 8 (Curve 1). The examination of this curve along with the regression line demonstrates that both the central value (ρ = 1.44 when C = 1mg/m^3) and the somewhat mysterious slope (-0.55) are in remarkable agreement within the middle range. The expression can be made more realistic by setting $\overline{b}_b(560)$ = 2% for C < 0.1mg/m^3 and by allowing it to vary from 2% to 0.5% when C goes from 0.1 to 20 mg/m^3, since it was recognized that algal cells exhibit low backscattering efficiencies (Morel and Bricaud, 1981). This

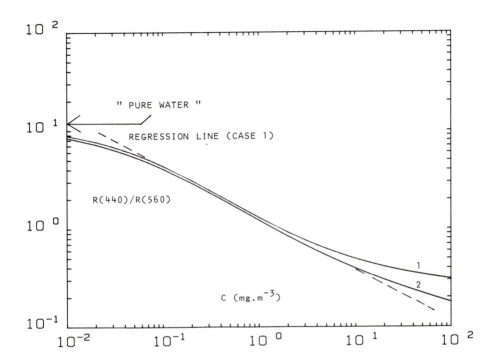

Figure 8. The dashed line represents the regression $\rho = 1.437C^{-0.556}$ (i.e., the inverted algorithm 3, Table 2). The upper and lower curves represent the variations of ρ according to the prediction of improved models (see text) with respect to models previously used in Fig. 4. The arrow indicates the limiting value $\rho = 11.9$ for pure water (when C tends to zero).

theoretical study also demonstrated the importance of the depressive effect of absorption on scattering, an effect which is clearly evidenced by in—situ determination of $b_b(\lambda)$ (Morel and Prieur, 1975). Accordingly, a tentative value $\overline{b}_b(440) = 0.3\%$ for $C = 20$ mg/m^3 is adopted. At low concentration, i.e., $C < 0.1$ mg/m^3, a wavelength dependence (λ^{-1}) is introduced for b_b which requires $\overline{b}_b(440) = 2.54\%$ in this domain. With these tentative values for low and high C, the variations in $\overline{b}_b$ at 440 and 560 nm are given by

$$\overline{b}_b(440) = 1.005 \ C^{-0.404}$$

and

$$\overline{b}_b(560) = 1.009 \ C^{-0.262}.$$

When these refinements are introduced, the result is

$$\rho = \frac{17.95 + 21.6C^{0.22} + 9.74C^{0.64} + 11.7C^{0.86}}{1.51 + 5.08C^{0.36} + 9.27C^{0.71} + 31.2C^{1.07}},$$

which, as expected, accounts more adequately for the low ρ values observed in the range of high $(C > 3$ mg/m$^3)$ concentrations (See Figure 8, Curve 2). The virtue of these relationships compared with the power law regressions is that they possess realistic limiting values when C tends toward very low or very high values, i.e., when the regression line fails and obviously becomes meaningless.

4. Preliminary Conclusions

Finally, with the proviso of considering only Case 1 waters, a set of mutually consistent relationships exists which cannot be other than an expression of natural laws summarized below:

i) Associated detrital material is always present along with living phytoplankton. However, the relative proportions of these by-products and of living cells apparently vary in a continuous manner, with a relative decreasing influence of detritus when the pigment concentration increases. That leads to a nonlinear biological effect which is reflected by the exponents of the order 0.6 or 0.7 appearing in the C-S, as well as in the K-C relationships. (Note that the particulate organic carbon (POC)-chlorophyll $\underline{a}$ relationship shows the same trend (see, e.g., Hobson et al., 1973). It is not really surprising, when considering that high pigment concentrations are found in nutrient-rich upwelled waters supporting vigorously growing algal populations. After some delay the grazing pressure increases, creates debris, and reduces the algal standing crop.)

ii) With such a continuous law governing the proportions between debris and living cells, and to the extent that the optical properties of both these components can be regarded as constant, a corollary proposition can be derived. A unequivocal relationship must rule the change in the spectral behavior of K (K_d or K_u) with changing C, and also the spectral shape of upwelling radiance (for given conditions of incident radiation).

Interpretation of Remotely Sensed Ocean Color

Now, the highest correlation coefficients seem to be found for the $K-r_{ij}$ relationships (see Table 2). Within the restriction of dealing only with Case 1 waters, a simple way of using the remote measurements of ocean color could be to derive an index of the bio-optical state of the water. This index conveniently should be K, or more precisely $i_\lambda = (K-K_w)_\lambda$. From this index, and as by-products, C and S could be straightforwardly derived and scales of equivalence be constructed.

According to this suggested scheme, a coherent set of interrelationships must be adopted. It is assumed that the indices i_{490} or i_{520} have been obtained through the Austin-Petzold algorithms (9) and (10). (Note that if $L_u(443)$ has become undetectable in case of high C concentration, the relationships between spectral upwelling radiances derived by these authors allow other radiance ratios to be used, at least in principle. With $L_u(520)$ replacing $L_u(443)$, for instance, the following algorithms can be obtained:

$$i_{490} = 0.162[L_u(520)/L_u(550)]^{-5.78};$$

$$i_{520} = 0.117[L_u(520)/L_u(550)]^{-5.42}.$$

They are, as expected, less sensitive than (9) and (10) owing to the high negative values in the exponent.)

The second step consists of calculating C and S. The relationships (19), slightly rounded,

$$i_{490} \cong 0.07C^{0.70} \qquad\qquad i_{520} \cong 0.05C^{0.70},$$

combined with (14) lead to the following set of equations:

$$C = 44.6(i_{490})^{1.43} \quad \text{or} \quad C = 72.2(i_{520})^{1.43}$$

$$S = 3.16(i_{490})^{0.89} \quad \text{or} \quad S = 4.26(i_{520})^{0.89}.$$

These relationships are displayed in Figure 9. It is estimated that C and S can be derived from i_λ to within a factor of two (i.e., ±40%) at most, in the full range of concentration expected at sea. In the diverse regression analyses concerning the various parameters, the standard error of estimate has approximately the same value which is at the origin of the factor two.

Since only Case 1 waters are under consideration, such a factor may appear rather high and the question arises: is an improved accuracy achievable, for example as a result of statistical analysis covering a much larger body of data? It seems doubtful that considerable improvement should be expected (except maybe in case of local studies, using adapted algorithms) for ubiquitous Case 1 algorithms because natural sources of noise affect the interrelationships. The problems of pigment measurements and of coupling optical measurements with a representative and simultaneous set of pigment values for the water column lead to experimental errors which also contribute to the inaccuracies in the algorithms. These problems, however, are left out of the following discussion as an experimental artifact. Natural causes of the deviation are briefly recalled:

i) with respect to the average law governing the proportions between living cells and associated debris, fluctuations occur which obviously impair the tightness of all the other relations;

ii) the specific absorption of phytoplankton (i.e., per unit of chlorophyll a concentration) is actually

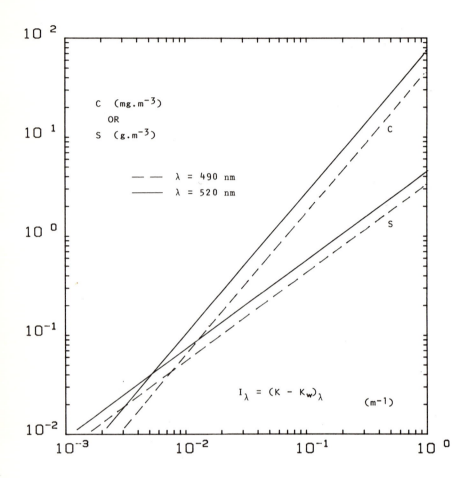

Figure 9. Curves giving the pigment or the seston concentration as a function of the bio-optical state indices i_λ, for $\lambda = 490$ and 520 nm.

variable with algal species according to the
size of the cells and their intracellular
pigment concentration (the 'discreteness
effect', Morel and Bricaud, 1981b);

iii) the shift in ocean color – from blue
to green with increasing biomass – which
finally is the base for estimating C is
caused by the 'blue absorption' due to all
plant pigments (and phaeopigments).
Chlorophyll a, indeed, is responsible for
the presence of a peak at 440 nm (Soret
band) but not for the whole broad absorption
in this spectral region where carotenoids
and accessory pigments play a major, even
dominant role (see, e.g., Figure 1a in
Bricaud and Morel,1981). The fact that
these accessory pigments covary reasonably
well with C enables its estimation from
water color measurements of this blue
absorption. However, this covariation is
not perfect and the ratio of blue absorption
to the chlorophyll a plus phaeophytin a
concentration is varying from one species to
another because of the change in accessory
pigments composition. (As a matter of fact,
the conventional spectrophotometric method
for the chlorophyll a determination makes
use of the specific red, not the blue
absorption peak). Consequently, the index
C, which is expressed in terms of the
chlorophyll a plus phaeophytin a
concentration, being derived from blue
absorption, is necessarily imperfect. The
bio-optical indices, as suggested above, are
from this point of view more reliable.

Considering the above natural sources of 'noise'
in the algorithms, the fact that the error is as low
as ±40% (and in fact lower if smaller ranges in C are
considered) is indeed remarkable.

IV. ATMOSPHERIC CORRECTION

There is little doubt that the atmospheric correction algorithm described by Eq. 13 in Section II-B is applicable over some portion of a CZCS scene when $S(\lambda, \lambda_0)$ can be found, and when $L_w(\lambda_0) = 0$. Initially, it was envisioned that $S(\lambda, \lambda_0)$ could be determined from surface measurements of the aerosol optical depth τ_A, using Eq. 12 in conjunction with the assumption that ω_A and P_A are independent of wavelength. Alternatively, measurements of L_w (actually L_u) could be used directly with Eq. 13 to provide $S(\lambda, \lambda_0)$. Since it is unlikely that $S(\lambda, \lambda_0)$ would be determined at many locations in an image, one is actually forced to assume that $S(\lambda, \lambda_0)$ remains essentially constant even though the turbidity of the atmosphere may vary considerably from location to location. Gordon (1981a) has shown that $S(\lambda, \lambda_0)$ can in fact remain nearly constant in a horizontally inhomogeneous atmosphere. As mentioned above, a sufficient condition for this is that the aerosol size distribution (normalized to the total concentration) and refractive index be independent of position over the image even in the presence of variations in the total concentration. As long as the aerosol type (i.e., continental or Marine) remains constant, this condition is probably reasonably well satisfied. Unfortunately, the data also suggest that the actual values of $S(\lambda, \lambda_0)$ derived from Eq. 13 do not conform well to what would be expected on the basis of Eq. 12. It is possible that this is due to spectral variations in either ω_A or the aerosol scattering phase function, which would cast doubt on the usefulness of surface

measurements of τ_A in estimating $S(\lambda, \lambda_0)$.
However, Gordon (1981a and 1981b) has shown that a
combination of uncertainty in the extraterrestrial
solar irradiance and small errors in the sensor
calibration (well within the estimated calibration
accuracy) could result in very large uncertainties in
$S(\lambda, \lambda_0)$, particularly in relatively clear
atmospheres. This is due to the fact that computation
of L_R in Eq. 13 requires the extraterrestrial solar
irradiance F_s, while computation of L_T requires
sensor calibration. Because L_R represents a
significant portion of L_T (typically ~ 80% in the
blue band), and the errors in the sensor calibration
and F_s are completely independent, large errors are
possible in the difference $L_T - L_R$. Such errors
could explain the sometimes erratic behavior in
$S(\lambda, \lambda_0)$.

   With the uncertainties inherent in applying Eq.
13 using surface measurements of τ_A, the
determination of $S(\lambda, \lambda_0)$ would seem to require
that either L_W be measured at one position in the
image or that 'clear water' (low pigment
concentrations) areas in the image be located. For
these areas L_W is then estimated from measurements
for similar solar zenith angles and expected pigment
concentrations. This requires the establishment of a
'clear water spectral radiance' data base and an
oceanographic knowledge of areas where such waters are
presumably present. (This concept is discussed further
in Appendix II.)

   As mentioned in Section II-B, the requirement
that there exist an available spectral band (λ_0)
for which $L_W = 0$ places a restriction on the waters
for which Eq. 13 may be applied. For example, in
coastal regions with high sediment concentrations it
is in general not possible to find such a λ_0. Two
methods have thus far been employed in an attempt to
relax this requirement. In the first method, which
Sturm (1981) has applied to very turbid water, it is
assumed that for some pixels in the scene L_W (with
$\lambda_0 = 670nm$) is actually zero. The aerosol
radiance is then computed for these 'darkest pixels'
in the scene. For all other pixels, L_A is evaluated

by transformation from the viewing angle for the 'darkest pixel' to that of the pixel under consideration and to the other spectral bands through the use of an aerosol model. This technique assumes that τ_A is spatially constant, a much more severe restriction than requiring that $S(\lambda,\lambda_0)$ be spatially constant. The resulting correction is, of course, model dependent.

In the second method, developed by Smith and Wilson (1981), an attempt is made to determine the water-leaving radiance at λ_0 (L_w) directly from the CZCS by employing an iterative procedure on a pixel by pixel basis. This is initiated by finding L_w from Eq. 13 with $\lambda_0 = 670$nm, i.e., assuming $L_w(\lambda_0) = 0$. Then, a relationship

$$(20) \quad L_w(670) = 0.083 L_w(443)[L_w(443)/L_w(550)]^{-1.661}$$

determined by Austin and Petzold (1981) is used to estimate $L_w(670)$ from L_w. This new estimate is then used in

$$(13') \quad tL_w = L_T - L_R$$

$$- S(\lambda,670)[L_T(670) - L_R(670) - tL_w(670)]$$

(Eq. 13 written to include $L_w(670)$ explicitly) to determine a new set of inherent sea surface radiances. These are again inserted into Eq. 20 and the process continued until convergence is achieved. In order to carry out this process, it is necessary to know $S(\lambda,\lambda_0)$. Smith and Wilson determine S by using ship measurements of τ_A in Eq. 12, assuming a spectrally independent single scattering albedo and phase function. Considering the difficulties with Eq. 12 mentioned above, this part of their procedure is open to question; however, the excellent agreement achieved between ship measurements and CZCS

determinations of pigment concentrations in the
Southern California Bight suggests that the basic
iteration technique is sound. As already pointed out
in Section III-E, an equation of the form of Eq. 20
implies the existence of a unique relationship between
the spectral shape of the inherent sea surface
radiance and the pigment concentrations. This may be
nearly realized for Case 1 waters, but is not likely
to be true for Case 2 due to the diversity in the
nature of their suspended materials. Thus, for waters
heavily loaded with inorganic sediment (such as are
found, for example, near estuaries as the Mississippi
Delta, or in the North Sea, or along the coastline in
the Mauritanian upwelling area, or in waters
influenced by glacial runoff), it may be necessary to
determine site-specific values for these coefficients
from preliminary optical field measurements.

V. APPLICATION OF THE ALGORITHMS TO CZCS IMAGERY

The initial application of the atmospheric
correction algorithm to CZCS imagery made by Gordon,
Mueller and Wrigley (1979) suggested that atmospheric
effects could be removed even in a horizontally
inhomogeneous atmosphere. Following this Gordon et
al. (1980) applied both the atmospheric correction and
in – water algorithms to imagery from the Gulf of
Mexico. Figure 10 shows a subscene of CZCS imagery at
443 nm from the Gulf of Mexico before and after
removal of the effects of the atmosphere. Orbit 130
is an example of a turbid atmosphere with aerosol
optical thickness $\tau_A(670)$ varying between 0.16 and
0.25. Note the strong horizontal inhomogeneities in
the haze which have been successfully removed by the
correction algorithm. Orbit 296, on the other hand, is
more representative of typical atmospheric conditions
for the region ($\tau_A(670) = 0.07$), and some water
structure can be seen even without atmospheric
removal. Application of the pigment algorithms (Gordon
and Clark, 1980b) to the atmospherically corrected
subscene from Orbit 296 is presented in Figure 11
(following page 90), in which the left panel gives the
pigment estimation based on $L_w(443)/L_w(550)$, while
the right panel estimation is based on
$L_w(520)/L_w(550)$. Two algorithms were used because
$L_w(443)$ decreases rapidly with pigment concentration
and for $C > 1mg/m^3$) typically becomes too small to
extract with confidence from the atmospherically
generated radiance. Near this concentration it
becomes necessary to switch to an algorithm which does
not involve 443 nm. The solid line on the image is
the ship track of the R.V. Athena II from 1500 hrs on
November 13 to 2000 hrs on November 14, 1978. The

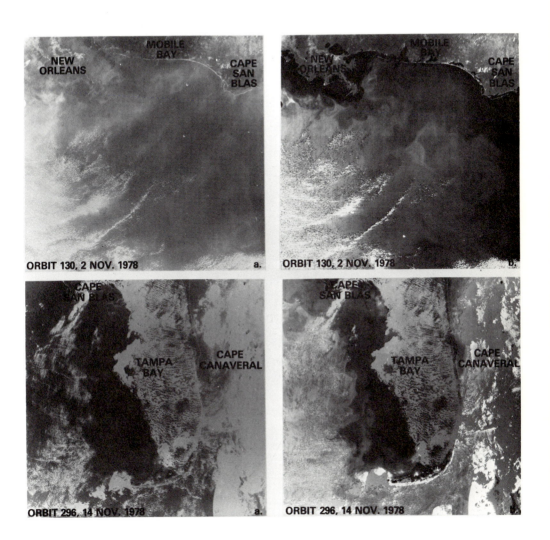

Figure 10. Comparison of CZCS-measured radiances in the blue band (a) and the atmospherically corrected radiance $L_w(443)$ (b) for Orbits 130 and 296 over the Gulf of Mexico. (Reprinted from Gordon et al., 1980. Copyright 1980 by the American Association for the Advancement of Science.)

satellite overpass was at about 1200 hrs on November
14, at which time the ship was occupying a station
west of Tampa. The surface determination and CZCS
estimation of the pigment concentration using the
$L_w(520)/L_w(550)$ ratio are compared in Figure 12.
The rms error in C is seen to be roughly ± a factor
of two. During the occupation of the station (6
hours) a chlorophyll variation of a factor of two was
also observed, which underscores the difficulty
encountered in comparing two data sets, one obtained
over a period of 29 hours, and the other over a period
of only one minute. In Figure 13, the same
ship — CZCS comparison has been reproduced using
Clark's (1981) improved algorithm (Table 2 (7)). Note
the apparent improvement in the comparison between the
two data sets.

    In Figure 14 an image of the Southern California
Bight which was processed at the Visibility Laboratory
using the iterative algorithm of atmospheric
correction (Smith and Wilson, 1981) is reproduced.
The shipboard and satellite determined pigment
concentrations for the track shown on Figure 14 are
compared in Figure 15. (C_{13}) SAT is the CZCS
estimated radiance using the ratio $L_w(443)/L_w(550)$
and the algorithm coefficients given in Table 2 (4).
The agreement between the two data sets is again seen
to be within a factor of 2. During the period 0800 to
1600 on 27 February the ship was occupying a station,
continuously recording fluorometrically determined
chlorophyll—like pigments. A factor of 2 variation in
C is clearly evident in the surface data during this
time, and Smith and Wilson suggest the possibility
that some of the disagreement between ship and
satellite determinations is due to real variations in
C during the acquisition of the two data sets. It is
interesting to note that, using this atmospheric
correction technique, Smith and Wilson appear to be
able to achieve excellent results using $L_w(443)$ even
at very high (up to $5mg/m^3$) pigment concentrations.
This may indicate a very high accuracy for their
correction technique or may be due to an interaction
between the atmospheric correction and in — water
pigment algorithm, both of which use in — water

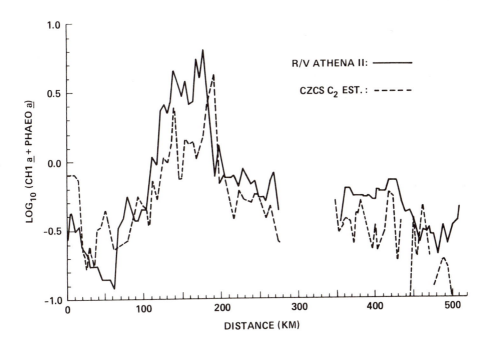

Figure 12.   Comparison between the CZCS-derived and ship-measured pigment concentration for the ship track in Figure 11.   (Reprinted from Gordon et al., 1980. Copyright 1980 by the American Association for the Advancement of Science.)

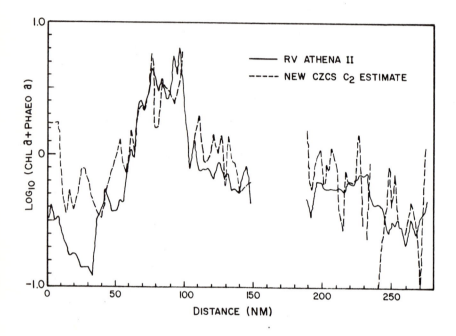

Figure 13.   Comparison between the CZCS-derived and
ship-measured pigment concentration for the ship track
in Figure 11 using Clark's (1981) improved pigment
algorithm (Table 2, algorithm 7).

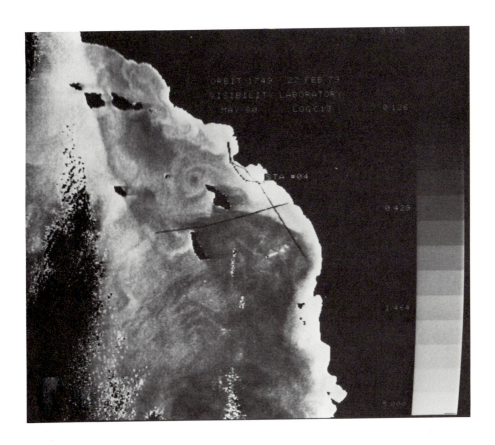

Figure 14.   Image of the pigment concentration in the Southern California Bight, derived using the Smith and Wilson (1981) iterative algorithm for atmospheric correction. (Reprinted from Smith and Wilson, 1981.)

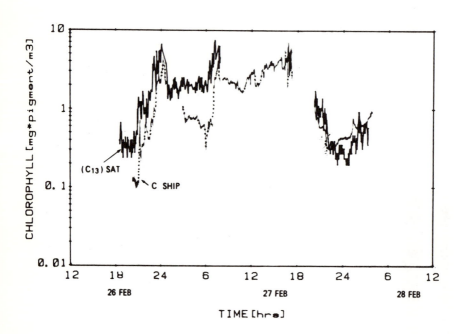

Figure 15. Comparison between the ship-measured and CZCS-derived pigment concentration for the image shown in Figure 14. The pigment algorithm employed here is algorithm 4 in Table 2. (Reprinted from Smith and Wilson, 1981.)

radiance data. It is felt that this merits further
study.

One interesting point which has emerged from
imagery studied thus far is the observation that the
visible imagery and thermal imagery are not redundant.
While most frontal features observed in color imagery
are coincident with thermal fronts, there are striking
exceptions. Indeed, Mueller and LaViolette (1981)
present an interesting image of the Gulf Stream's
north wall, obtained over the Grand Banks, for which
the sharpest thermal front is displaced over about 50
km from the strongest color front. In this case the
color front associated with the north wall was south
of the Gulf Stream's thermal boundary. A possible
explanation of this observation is that a thin layer
of warm clear Gulf Stream water was overlying the
cold-turbid-productive waters north of the stream.
The thermal signature, being indicative of the
temperature of the top 0.05 mm of the water, would
then indicate Gulf Stream water. On the other hand,
the color signature resulting from a weighted average
of optical properties over the top several meters
would be strongly influenced by the turbid water
below, until the layer thickness increased beyond the
clear water penetration depth. Nearly simultaneous
AXBT data across the thermal front, however, clearly
indicate that this explanation is not correct.
Mueller and LaViolette suggest the alternative
explanations of horizontal cross-front diffusion, or
the enhanced productivity resulting from weak
upwelling induced by secondary circulations in the
vicinity of the thermal front.

This section reflects the state-of-the-art at the
time of the Symposium in Venice (1980), when very few
images had been processed and fully interpreted.
Recent developments are the topic of Appendix II.

VI. SUMMARY AND CONCLUSIONS

The initial problem addressed in ocean color remote sensing is the determination of the concentrations of various ocean constituents (phytoplankton, total seston, etc.) in surface waters through measurement of the spectral radiance $L_T(\lambda)$ leaving the Earth's atmosphere. The physics of this problem is well understood. That is, the constituents influence the ocean's optical properties ($a(\lambda)$ and $\beta(\gamma,\lambda)$), and given the optical properties of the ocean and the atmosphere (and their distribution, respectively, with depth and altitude), and other environmental factors such as surface wind velocity, the spectral radiance $L_T(\lambda)$ at the sensor can be determined by solving the radiative transfer equation. Solving the inverse problem, the determination of the optical properties $a(\lambda)$ and $\beta(\gamma,\lambda)$ and hence the constituent concentrations, from measurements of $L_T(\lambda)$ is impossible without the addition of some simplifying facts or assumptions. One considerable simplification is the decoupling of the atmospheric and oceanic parts of the problem, and the fact that the ocean's contribution to $L_T(\lambda)$ to a good approximation depends only on the irradiance ratio $R(\lambda)$. However, even the inversion of $R(\lambda)$ to provide constituent concentrations requires extensive detail concerning the optical properties of the constituents, and this 'analytic' algorithm (see above, Section III-D) can be applied only to areas for which such information is available. Fortunately, most of the oceanic waters as a rule belong to Case 1, i.e., the variation of their optical properties from those of pure sea water is due chiefly to the presence of phytoplankton and their associated detrital material. The optical properties of such waters have been expressed in terms of a single parameter, the sum of the concentrations of chlorophyll _a_ plus phaeophytin _a_, C, called the pigment concentration. The first algorithms to extract C from $R(\lambda)$ related ratios of R or L_w at two wavelengths (usually 440 and 550 or 560 nm, where phytoplankton pigments are, respectively, strongly and weakly absorbing) to C. Separate algorithms of the same form were developed to extract the diffuse attenuation coefficient K, and the total seston, S. The interrelation between these various algorithms (culminating in a 'derivation' of the C algorithm from those for K and S) is elucidated in detail in Section III-E above, and is considered to constitute proof that: i) the optical properties of oceanic water are dominated by phytoplankton and their associated detrital material; and ii) a 'universal,' albeit noisy, pigment algorithm exists for such oceanic waters (see above, Section III-E, for a discussion of the various noise sources).

For waters near the coast, which often do not fall into the above category and are referred to as Case 2 (see above, Fig. 2), the situation is more complex. The optical properties then usually depend on the concentration of several (independent) constituents. In certain cases, e.g., the Coastal U.S., pigment algorithms have been developed which enable the retrieval of C from $L_w(\lambda)$ with an accuracy of about ±70–80% (compared to ±30–40% for Case 1 waters). One expects, however, that such Case 2 water pigment algorithms will be very site-specific.

Before these 'in-water' algorithms can be applied the effect of the atmosphere on L_T (L_1 in Eq. 11) must be removed. Again, given the optical properties of the atmosphere, computation of L_1 is straightforward. However, in practice these are unknown and L_1 must be determined from $L_T(\lambda)$ alone. This is accomplished by: first, removing the effect of Rayleigh scattering (L_R in Eq. 11); next, computing the aerosol contribution L_A using a spectral band, λ_0 (usually 670 nm), at which the water is black ($L_w(\lambda_0) \cong 0$); and, finally, determining the spectral behavior of the aerosol

radiance $L_A(\lambda)$, by examining a Case 1 region of
the image where the pigment concentration is less than
0.25 mg/m^3 (see below, Appendix II). This spectral
behavior is then assumed to apply to the entire image,
providing $L_A(\lambda)$ for each pixel which satisfies the
condition $L_w(\lambda_o) \cong 0$. At sufficiently large C,
even in Case 1 waters, there is no such λ_o for the
CZCS. This problem is circumvented for Case 1 waters
by the existence of a (nonlinear) relationship between
L_w's (see, e.g., Eq. 20) which may be used to
replace the equation $L_w(\lambda_o) = 0$. However, when
no λ_o exists for Case 2 waters, a site-specific
relationship among the L_w's (to replace Eq. 20) must
be established in order to effect the atmospheric
correction.

   The initial application of the algorithms to CZCS
imagery suggested that for Case 1 waters C could be
retrieved from $L_T(\lambda)$ at least to within a factor
of ±2. More recent validation exercises using the
improved algorithms (post-launch algorithms (5) and
(7) in Table 2, or an algorithm restricted to Case 1
waters, Appendix II), or algorithms that are
'calibrated' using simultaneous ship measurements,
indicate that C can be retrieved to within ±35-40%,
i.e., with an error only slightly greater than that
produced by the natural 'noise' in the pigment
algorithms. Also, atmospheric correction can now be
accomplished without recourse to any surface
measurements.

   Some of the oceanic imagery processed thus far
clearly indicates that color and thermal imagery are
not redundant. In general all thermal features (weak
and strong fronts) tend to have associated color
features. The converse, however, is not true; color
features exist without accompanying thermal structure.
Thus, the color imagery appears to be richer in
structure than the thermal imagery, and along with its
initial goal (the estimate of algae and sediment
concentration) this imagery in the near future will
most likely constitute a remarkable tool for the
spatial-temporal description of structures at small
and intermediate scales in the oceans. This is in
part due to the ability of a short wave sensor to

'look' into the sea and hence to distinguish between water masses which are hardly distinguishable by their 'skin' temperatures; however, the main ability of this tool rests on the reasonable assumption that phytoplankton and their derivatives are passive tracers of turbulence. (Gower et al., 1980, have published a spectral analysis of spatial structures (attributed to phytoplankton) from Landsat imagery of waters just to the south of Iceland, and point out that their fluctuation spectrum is consistent with the phytoplankton behaving as a passive scalar, an assumption to be used with some caution (Gower and Denman, 1981; Lesieur and Sadourny, 1981).)

It seems reasonable to conclude that ocean color remote sensing should now be considered well understood. The subject rests on a firm physical foundation, and the procedures described in the text can be applied with confidence to Case 1 waters, for which 'universal' algorithms apparently exist. For Case 2 waters there is no conceptual difference; the situation is simply made more complex by virtue of the diversity of possible materials in the water, necessitating site-specific algorithms. Although inconvenient, this should not be considered a shortcoming of the techniques. Ocean remote sensing should be judged on its capacity to provide important auxiliary information for the study of oceanographic problems, not on its capacity for solving problems alone.

VII. APPENDIX I: THE COASTAL ZONE COLOR
SCANNER (CZCS)

The Coastal Zone Color Scanner (CZCS) is a
scanning radiometer which views the ocean in six
coregistered spectral bands, five in visible and near
infrared (443, 520, 550, 670, and 750 nm), and the
sixth, a thermal infrared band (10.5-12.5 μm). The
satellite is in a sun-synchronous orbit with ascending
node near local noon. The scanner has an active scan
of 78° centered on nadir, and a field of view of
0.0495°, which from a nominal height of 955 km,
produces a ground resolution of 825 m. The scanning
is accomplished through 360° turns of a rotating mirror.
When the sensor is viewing at nadir the normal to the
mirror lies in the plane formed by nadir and the
direction of motion of the spacecraft and makes an
angle of 45° with each. To minimize the effect of
direct sun glint the sensor is equipped with a
provision for tilting the scan mirror in such a manner
that the angle the mirror normal makes with nadir can
be varied by ± 10° in 1° increments. This has the
effect of tilting the scan plane through ± 20° in
2° increments. A secondary effect of this tilting is
a widening of the 1600 km swath to 2300 km when the
scan plane is tilted north by 20°, and a narrowing
of the swath to 1300 km when the scan plane is tilted
20° to the south.

Radiometric calibration is achieved by viewing
one of two calibration lamps every 16 scan lines. At
typical signal levels the signal-to-noise ratio is
above 150 for all bands with the exception of band 4,
for which the measured value is 118 (Hovis et al.,
1980).

As discussed in the text, the purpose of the CZCS

experiment is to provide estimates of the near-surface
concentration of phytoplankton pigments and total
seston by measuring the spectral radiance
backscattered out of the ocean. Since this radiance
is nearly always considerably less than 1
mW/cm^{2}μm ster (except in the blue for very clear
water) the sensor must possess very high radiometric
sensitivity. Table A I-1 gives the spectral bands and
saturation radiances for the CZCS shortwave bands at
Gains 1 and 4. Included for comparison are the
Landsat Multi-Spectral Scanner (MSS) bands within
which the CZCS bands are contained, and the MSS
saturation radiances. At sun angles typical with the
MSS, the CZCS would normally be operated at Gain 4 so
the radiometric sensitivity of the CZCS is roughly an
order of magnitude greater than that of the MSS. Also,
the CZCS is 8-bit digitized on board compared to 6-bit
for the MSS, providing another factor of four in
radiometric sensitivity.

Table A I-1: Comparison of NIMBUS-7 CZCS and LANDSAT-1 MSS spectral and radiometric sensitivities (λ is in nm and the saturation radiance, S.R., is in mW/cm^2 μm ster.).

	CZCS				MSS	
Band	λ	S.R. Gain 1	Gain 4	Band	λ	S.R.
1	433–453	11.46	5.41			
2	510–530	7.64	3.50	5	500–600	24.8
3	540–560	6.21	2.86	5	500–600	24.8
4	660–680	2.88	1.34	6	600–700	20.0
5	700–800	23.90	23.90	7	700–800	17.6

VIII. APPENDIX II: RECENT DEVELOPMENTS

This appendix summarizes the major developments in ocean color remote sensing which have taken place between the COSPAR/SCOR/IUCRM Symposium and the IAMAP Assembly in Hamburg (August 1981).

A. Clear Water Radiance Concept

As suggested in Section IV, a possible scheme for computing the atmospheric correction factors $S(\lambda, \lambda_o)$ is to examine a position in the image at which the pigment concentration is sufficiently low so that L_w can be reliably estimated. Gordon and Clark (1981) have demonstrated that L_w can in fact be estimated at 520, 550, and 670 nm to within 10% for Case 1 waters with $C < 0.25$ mg/m^3. From the relationships developed in Section II-A it is easy to show that

$$L_w = E_d(0)(1-\rho)R/Qm^2,$$

where ρ is the Fresnel reflectance (water to air) of the interface. Also, the downwelling irradiance just beneath the surface $E_d(0)$ is related to the extraterrestrial solar irradiance by

$$E_d(0) = (1-\bar{\rho})\mu_o F_s t,$$

‾ is the albedo of the sea surface
ing irradiance scattered out of the water) and
given by Equation 10 with μ replaced by μ_0.
The influence of the atmosphere and the solar zenith
angle on L_w is completely determined by their
influence on $E_d(0)$. This suggests that these
influences can be removed from L_w by forming

$$L_w = (L_w)\mu_0 t,$$

where the normalized water-leaving radiance (L_w) is
given by

$$(L_w) = \left[(1-\rho)(1-\overline{\rho})F_s/Qm^2 \right] R.$$

In this expression only R depends on the pigment
concentration. Spectra of $R(\lambda)$ for Morel's Case 1
waters (Figure AII-1) show that for $\lambda > 500$ nm, R
(and hence L_w) is nearly independent of C if
$C < 0.25$ mg/m^3. Thus, it is to be expected that
(L_w) will be nearly independent of C for
$C < 0.25$ mg/m^3. In principle (L_w) can be found
from surface measurements of L_w only if the term
$(1-\omega_A F)\tau_A$ in the above equation is known. In
practice, however, this term usually can be ignored
since $(1-\omega_A F)$ is typically less than 1/6 for most
aerosol models. From the analysis of 11 stations in
the Gulf of Mexico and Sargasso Sea, for which C was
less than 0.25 mg/m^3, Gordon and Clark conclude that

$$(L_w(520)) = 0.495 \pm 7\%,$$

$$(L_w(550)) = 0.280 \pm 8\%,$$

and

$$(L_w(670)) < 0.012 \text{ mW/cm}^2 \text{ } \mu m \text{ ster.}$$

The values of $(L_w(520))/(L_w(550))$ computed from these radiances are in excellent agreement with

$$\left[R(520)/R(550) \right]\left[E_d(520)/E_d(550) \right]$$

(where $E_d(\lambda)$ is the downwelling irradiance just above the surface) computed from the Morel and Prieur data presented in Figure AII-1.

These radiances can be used to determine $S(520,670)$, $S(550,670)$, and $S(670,670)$ (=1) by examination of a clear water calibration area ($C < 0.25$ mg/m^3) from which $S(440,670)$ can be estimated by extrapolation. The best clear water calibration area in an image can usually be found by first making an approximate atmospheric correction using

$$S(\lambda,\lambda_o) = F_o(\lambda)/F_o(\lambda_o),$$

then computing the pigment concentration using the $L_w(443)/L_w(550)$ ratio, and finally selecting regions in the scene which appear to have sufficiently low pigment concentrations to qualify as clear water, e.g., $C < 0.25$ mg/m^3. If several such regions are found in an image, that which yields the smallest values for $S(\lambda,\lambda_o)$ should be chosen to avoid over correcting the image.

It is obvious that determination of $S(443,670)$ from clear water areas by extrapolation requires that the dependence of $S(\lambda,\lambda_o)$ on λ be smooth and not erratic as was observed in the initial imagery (see above, Section IV). It is now believed that this erratic behavior of $S(\lambda,\lambda_o)$ with λ was due entirely to a lack of consistency between the sensor calibration and the extraterrestrial solar irradiance. Studies to determine empirical corrections to the

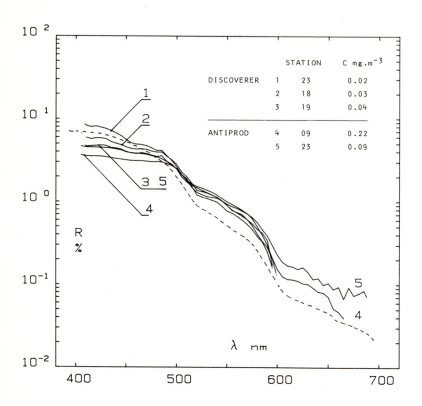

Figure AII-1. R(λ) for waters with low pigment concentration (Morel and Prieur, 1977a; Morel et al., 1978). Note the weak dependence of R(λ) on C, at these concentrations, for λ > 500 nm. The dashed curve represents the spectral reflectance computed for ideally pure water (curve labeled T1 on Figure 4 in Morel and Prieur, 1977a).

Figure AII-3. Pigment concentration derived using Clark's Case 1 algorithm (see text) applied to one and one-half CZCS images of the Western North Atlantic from Orbit 3226.

Figure AII-4. Same image as in Figure AII-3 with the pigment concentration derived from algorithm 7 in Table 2.

Figure AII-5. Output of the thermal sensor on the CZCS for the same image as in Figure AII-3. The temperature scale was estimated from ship measurements.

Figure AII-6. Pigment concentration derived from using Clark's Case 1 algorithm (see text) applied to a subscene of an image of the Mediterranean Sea south of Marseilles from Orbit 2090. The pigment scale is identical to that in Figure AII-3.

Figure 11. Ship track of the RV ATHENA II superimposed over an image of the derived pigment concentration from Orbit 296 using the prelaunch pigment algorithms of Gordon and Clark (1980b). (Reprinted from Gordon et al., 1980. Copyright 1980 by the American Association for the Advancement of Science.)

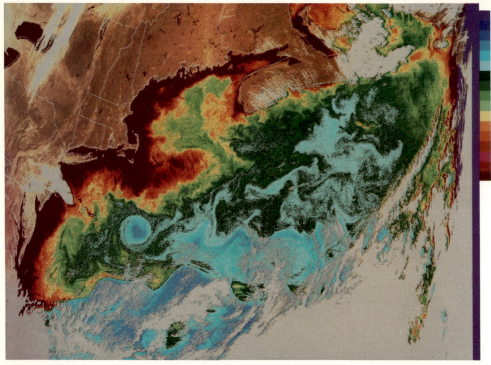

Figure AII-3.

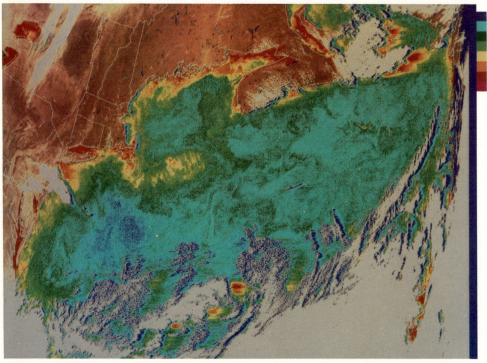

Figure AII-4.

Figure AII-5.

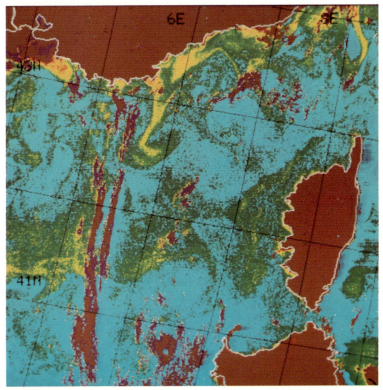

Figure AII-6.

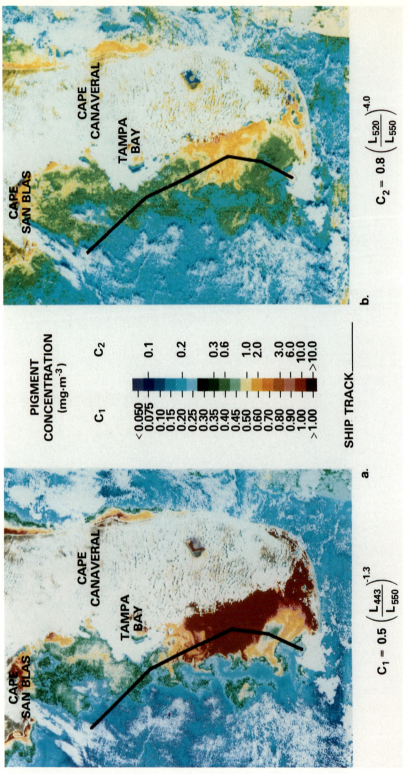

Figure 11.

sensor calibration to force such consistency have been
carried out (see below). This problem can be, and
should be circumvented on future systems through the
addition of a means to enable the sensor to view the
sun in diffuse reflection (Gordon, 1981b).

B.   Accuracy of Pigment Estimates

The above clear water radiance concept has been
applied by Gordon et al. (1983) to imagery of the
shelf and slope waters of the Middle Atlantic Bight
(most of which are close to Morel Case 1 waters), and
also to Sargasso Sea waters. Small adjustments in
sensor calibration were required to provide values of
$S(\lambda,\lambda_0)$ which were reasonably stable from day to
day, and not at strong variance in terms of their
dependence on λ. The radiance at the sensor is
written

$$L_T(\lambda) = \Big[A(\lambda)N(\lambda) + B(\lambda)\Big]C(\lambda),$$

where $A(\lambda)$ and $B(\lambda)$ are the calibration slope and
intercept, $N(\lambda)$ is the digital count from the
sensor, and $C(\lambda)$ is the calibration correction
resulting from the adjustment mentioned above. Using
the values of $A(\lambda)$ and $B(\lambda)$ given in
Table AII-1[*] along with the value of the mean
extraterrestrial solar irradiance (Table AII-2)
computed by weighing the Neckel and Labs (1981)
$F_s(\lambda)$ with the spectral response of the CZCS (R.W.
Austin, personal communication) yields the calibration
corrections $C(\lambda)$ listed in Table AII-1. It should

---

[*] These were provided by R.W. Austin, and have the
obviously desirable property that variation of the
sensor gain while viewing essentially homogeneous
water masses results in no variation in the computed
sensor radiance L_T.

Table AII–1. CZCS Calibration.

Band	A	B	C	GAIN
1	0.04452	0.03963	1.069	1
2	0.03103	0.06361	0.993	1
3	0.02467	0.07992	0.955	1
4	0.01136	0.01136	1.000	1
1	0.03598	0.05276	1.069	2
2	0.02493	0.08826	0.993	2
3	0.02015	0.06247	0.955	2
4	0.00897	0.03587	1.000	2
1	0.02968	0.02879	1.069	3
2	0.02032	0.09752	0.993	3
3	0.01643	0.06570	0.955	3
4	0.00741	0.02963	1.000	3
1	0.02113	0.03359	1.069	4
2	0.01486	0.05647	0.993	4
3	0.01181	0.04723	0.955	4
4	0.00535	0.01646	1.000	4

A and B are in mW/cm^2 µm sr

be noted that $C(\lambda)$ depends on the values used for $A(\lambda)$, $B(\lambda)$, and $\overline{F}_s(\lambda)$ (e.g., see Viollier, 1982). The values listed in Tables AII-1 and AII-2 are now being used by NASA to process CZCS imagery.

In all, five images were processed. The resulting CZCS-derived pigment concentrations computed using a C algorithm derived from Clark's data for Case 1 waters alone ($\lambda_i = 443$, $\lambda_j = 550$, $A = 1.13$ mg/m^3, and $B = -1.71$) were compared with continuous measurements made aboard ship within 12 hours of the satellite overpass. In all, a total of 569 individual comparisons were carried out with the result that over the range 0.08 to 1.5 mg/m^3 the pigment concentration was retrieved from the imagery with an accuracy of about 35%. This was particularly satisfying since the atmospheric turbidity (as indicated by the aerosol radiance at 670 nm) varied by a factor of 5 over the five orbits and showed considerable horizontal inhomogeneity in the individual images.

Table AII-2. Extraterrestrial Solar Irradiance $\overline{F}_s$.

Band	$\overline{F}_s$
1	186.42
2	185.34
3	184.76
4	151.52

$\overline{F}_s$ is in mW/cm^2 μm

In the same study, three direct comparisons between ship-measured and satellite-retrieved values of L_w were also made at locations 100–300 km from

the clear water calibration area. In these three cases
the atmospheric correction algorithm enabled retrieval
of the subsurface water radiance with an average error
of less than 10%, indicating that the assumption of a
constant $S(\lambda,\lambda_o)$ was valid on these three
occasions. Thus, the accuracy of the in-water
algorithm, and not atmospheric correction, was the
limiting factor for the accuracy with which pigment
concentrations could be retrieved in this study. It
is important to note again that this particular
atmospheric correction technique does not require any
surface measurements.

In a later paper (Gordon et al., in preparation)
these revised algorithms were applied to imagery
coincident with ship tracks having pigment
concentrations exceeding 1.5 mg/m^3. For this a
procedure similar to that used by Gordon et al. (1980)
was employed, i.e., different pigment algorithms were
used depending on the value of C. Specifically, if the
Case 1 algorithm described above yields
$C \leq 1.5$ mg/m^3, this is taken as the CZCS-derived
pigment concentration. If this algorithm yields
$C > 1.5$ mg/m^3, C is recomputed using an algorithm
similar to 7 in Table 2 but derived by excluding any
data for $C \leq 1.5$ mg/m^3 from the regression
(A = 3.23 mg/m^3 and B = −2.44). When this recomputed
C is also greater than 1.5 mg/m^3, it is taken to be
the CZCS-estimated concentration, while if the
recomputed C is less than 1.5 mg/m^3, the original
estimate derived from the Case 1 algorithm is used.
Figure AII-2 shows the result of applying this
procedure to the image from Orbit 296 shown in
Figure 11. The ship-measured pigment concentration is
the dashed line and the satellite-estimated
concentration is the solid line. Comparing Figures 12,
13, and AII-2 indicates significant improvement , with
time, in the CZCS retrievals. (Note: (1) zero on the
distance scale in Figures 12 and 13 corresponds
approximately to 220 km on the distance scale on
Figure AII-2; and (2) the pigment scale on Figures 12
and 13 is logarithmic, while on Figure AII-2 it is
linear.) This improvement is due to the application of
improved algorithms − both the atmospheric correction

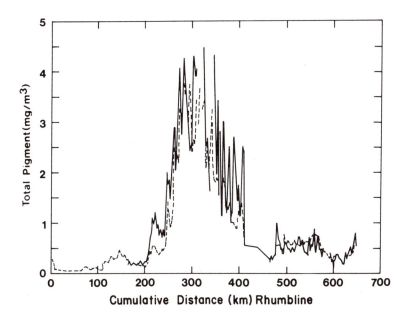

Figure AII-2.   Comparison between the ship-measured
and CZCS-derived pigment concentration for the image
shown in Figure 11. (See text for a description of the
algorithms used for this comparison.)

and the in-water pigment algorithms – and improvements in merging the ship navigation with geolocation of the imagery.

Using the Smith and Wilson (1981) iterative technique of atmospheric correction to obtain $L_w(\lambda)$, Smith and Baker (1982) have derived an algorithm specific to the Southern California Bight by fitting the pigment concentration measured along a ship track to CZCS estimates of r_{ij} ($\lambda_i = 443$, $\lambda_j = 550$, $A = 1.262$ mg/m^3, and $B = -2.589$). This enabled them to estimate pigment concentrations for the rest of the image to within an accuracy of $\pm 40\%$ over the range $0.05 < C < 5$ mg/m^3.

Using the same atmospheric correction technique, Austin (1981) has found excellent agreement between CZCS-estimated and surface-derived values of the diffuse attenuation coefficient K. He also demonstrated the existence, in the North Pacific, of a strong correlation between K measured over a depth 1/K and the average K over the upper 100m of the water, suggesting that at least for this area, CZCS K's are applicable to depths significantly greater than 1/K.

It should be noted that most atmospheric correction schemes discussed thus far employ the assumption that the correction factors $S(\lambda, \lambda_o)$ are position independent. However, for regions of an image for which it is known that $C < 0.25$ mg/m^3, e.g., the Sargasso Sea or much of the open ocean, this assumption can be relaxed (Gordon et al., 1983) by treating each pixel as a clear water calibration area. In this way $S(\lambda, \lambda_o)$ is recomputed at each pixel and the unknown $L_w(443)$ (and hence C) retrieved without any assumptions regarding the properties of the aerosol. Pigment retrievals in such cases are also found to agree with ship estimates to within about $\pm 35\%$.

C. Applications

Application of CZCS imagery to oceanographic problems has thus far been very limited due to the

paucity of images processed to 'derived products' C
and K. Investigators having access to
image–processing facilities have, for the most part,
been preoccupied with algorithm development and
'validation' studies, while those more interested in
applications generally have not had wide access to
image–processing facilities. Now, however, the
attention of investigators having access to such
facilities is beginning to focus on applications.
Gordon et al. (1982) have demonstrated that CZCS
imagery can be used quantitatively to observe the
surface pigment distribution in warm core Gulf Stream
rings (WCR) imbedded in the Slope and Shelf Waters of
the Middle Atlantic Bight. A striking example
(presented at the IAMAP Assembly (Gordon, 1981c)) of
an image containing a warm core ring is given in
Figure AII–3. This image (actually 1.5 CZCS images)
from Orbit 3226 on 14 June 1979, was atmospherically
corrected with the clear water radiance concept, and
the pigment concentration derived using Clark's Case 1
algorithm for 443 and 550 nm described above. The
accuracy in the pigment concentration is felt to be
±one color chip on the scale at the right. (The
meaning of the various colors for this and the rest of
the images in color is presented in Table AII–3.) The
WCR is the rich blue circle visible at a distance of
about 1/4 of the image south of Cape Cod. The
meandering north wall of the Gulf Stream is clearly
evident south of the WCR. Directly to the east of the
ring low pigment regions are seen which may be
indicative of other WCR's undergoing formation and
drifting toward the west. This image suggests that
the CZCS could be a valuable supplement to the study
of the mixing processes between the ring and the more
productive waters which they pass through.
  For comparison, the pigment concentration given
by an algorithm using 520 and 550 nm (algorithm 7,
Table 2) is presented in Figure AII–4. This algorithm
is useful for delineating areas of high pigment
concentration, for which $L_w(443)$ is too small to be
accurately retrieved from $L_T(443)$, but yields poorer
accuracies than the algorithms involving 443 nm.
Pigment structure is now evident on Georges Bank and

Table AII-3: Color scales for Figures A II-3, AII-4,
A II-5, and A II-6.

Number	Color	C(13) mg/m^3	T °C	C(23) mg/m^3
1	Purple	<0.05	—	——
2	Dk. Blue	0.075	6	0.10
3	Blue	0.10	8	——
4	Blue	0.15	10	——
5	Blue	0.20	12	0.20
6	Lt. Blue	0.25	14	——
7	Dk. Green	0.30	16	——
8	Green	0.35	18	0.30
9	Green	0.40	20	——
10	Lt. Green	0.45	22	——
11	Yellow	0.50	24	1.0
12	Lt. Orange	0.60	26	2.0
13	Dk. Orange	0.70	28	——
14	Rust	0.80	—	3.0
15	Pink	0.90	—	6.0
16	Red	1.00	—	10.0
17	Dk. Red	>1.00	—	>10.0

in the coastal waters using this algorithm. Note also that the WCR is still barely visible, which is remarkable considering that the contrast in C between the ring and the Slope Water would cause essentially no change in $L_T(520)$ and a change of only 1–2 digital counts in $L_T(550)$.

Figure AII–5 delineates the ocean 'skin' temperature as determined with the CZCS thermal band for the same scene as Figure AII–3. This thermal image has not been atmospherically corrected and in fact has not even been calibrated. It is actually an image of the thermal band digital counts ($\propto$ radiance); however, ship measurements of the SST (Clark, personal communication) made in this area nearly simultaneously with the image allow a coarse 'calibration' of the color scale to the right of the image (Table AII–3). The correspondence between the features in the pigment and SST imagery is striking. Such correspondence, however, is not universally observed.

Figure AII–6 shows the pigment concentration in the Mediterranean Sea south of Marseilles derived from Orbit 2090 (24 March 1979). (For a full scene of this image see Hovis, 1981.) This subscene was processed in a manner identical to that for the image in Figure AII–3. (The clear water region used for determination of the atmospheric correction parameters was located at 39.6°N and 6.09°E.) The CZCS-derived pigments agree well with measurements made both from ship and from aircraft flying at low altitude (Morel and Tailliez, unpublished). These measurements (22 March, 1979), starting from Monaco along a track headed 135° from North for about 25 km (ship), or 160 km (until reaching Corsica, aircraft) showed values in the range C < 0.2 to ~ 0.5 mg/m^3. Note the presence of considerable spatial structure in the pigment concentration. Figure AII–7 shows the same subscene imaged by the CZCS thermal band. The gray scale for this image has been stretched so that it contains only a portion of the full (0–255) count range: 145 (black) to 176 (white). Darker shades are warmer. Since no atmospheric corrections were attempted we cannot directly relate this image to the

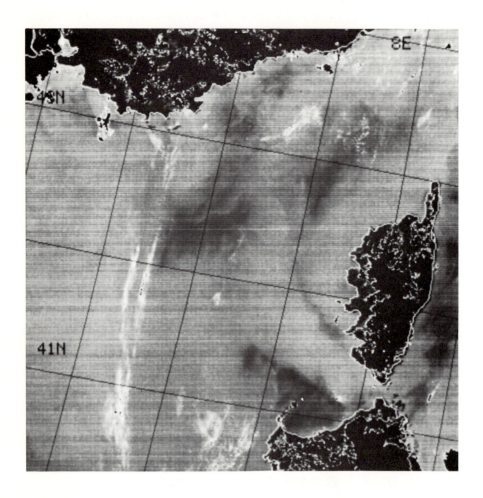

Figure AII-7. Contrast-stretched radiance from the CZCS thermal band for the same scene as presented in Figure AII-6. The water temperature variation over the entire scene is estimated to be approximately 2°C. Note the lack of correlation between the spatial structures in Figures AII-6 and AII-7.

sea surface temperature. However, we can use the sensor calibration to compute the temperature differences which would be present in the absence of atmospheric attenuation. For example, the temperature contrast between the warm water just north of Sardinia (near 41.3°N and 8.5°E) and the cooler surrounding waters is computed to be 1.6°C. Atmospheric effects influence satellite measurement of temperature contrasts $(\Delta T)_{Sat}$ according to

$$(\Delta T)_{True} = (\Delta T)_{Sat}/(Trans),$$

where (Trans), the atmospheric transmittance for the 10.5–12.5 μm range, is typically somewhat larger than 0.8 for this region and time (G. Maul, personal communication). Thus the temperature contrast computed above is probably no more than 2°C, and by the same reasoning the temperature variation over the entire subscene is only slightly larger than this.

The most striking aspect of Figure AII–7 is the presence of only weak spatial structure compared to that seen in the pigment concentration in Figure AII–7, and the apparent lack of correlation between the C and T images. This is in contrast to the strong correlation between Figures AII–3 and AII–5.

Other observations concerning the relationship between visible and thermal imagery have been made by several investigators. Wrigley (1980) has presented CZCS imagery from the central Pacific Ocean in the vicinity of 37° N and 170° W from early June to early July 1980. Thermal fronts with temperature differences of up to 2°C were observed in June; but by July their temperature contrast had weakened to 0.2°C. Evident in the imagery, however, were eddies and gyres with scales as large as 100 km which did not lose their color contrast with time. All the structures which were visible in the thermal imagery were also evident in the color imagery; however, the converse was not true. Gower (personal communication) has compared CZCS atmospherically corrected radiances in the visible with thermal imagery from the NOAA–6

AVHRR (more sensitive than the CZCS thermal sensor) off British Columbia, Canada, and finds considerable spatial structure in the visible which is not delineated in the IR. The inescapable conclusion of these studies is that visible and thermal imagery are not redundant.

Smith, Eppley, and Baker (1982) have compared single-position ship measurements of pigments with the areal distribution of C in the vicinity of the ship as inferred from the CZCS, to study the synopticity of ship measurements in the Southern California Bight. Their results indicate a marked asymmetry in the frequency distribution of C, such that the areal mean is usually larger than the most frequently occurring concentration. They have also derived a coarse relationship between C and primary productivity, and they conclude that the major difficulty in computing productivity from CZCS-derived pigments is the variance in this relationship.

IX. REFERENCES

Arvesen, J.C., Millard, J.P., and Weaver, E.C., 1973.
'Remote sensing of chlorophyll and temperature in
marine and fresh waters,' Astronaut. Acta., 18,
229-239.

Austin, R.W., 1974. 'The remote sensing of spectral
radiance from below the ocean surface, 'In Optical
Aspects of Oceanography, edited by N.G. Jerlov and
E.S. Nielsen, Academic Press, London. Ch. XIV,
317-344.

Austin, R.W., 1981. 'Remote sensing of the diffuse
attenuation coefficient of ocean water,' The 29th
Symposium of the AGARD Electromagnetic Wave
Propagation Panel on Special Topics in Optical
Propagation, Monterey, Calif., 6-10 April.

Austin, R.W., and Petzold, T.J., 1981. 'The
determination of the diffuse attenuation coefficient
of sea Water using the coastal zone color scanner,' in
'Oceanography from Space,' edited by J.R.F. Gower,
Plenum Press, New York, p. 239-256.

Banse, K., 1977. Determining the carbon-to-chlorophyll
ratio of natural phytoplankton, Mar. Biol., 41,
199-212.

Barton, E.D., Huyer, A., and Smith, R.L., 1977.
'Temporal variations observed in the hydrographic
regime near Cabo Corveiro in the N.W. African
upwelling region, February to April 1974.' Deep Sea
Res., 24, 7-23.

Bricaud, A. and Morel, A., 1981. 'Possible variations in the specific absorption by phytoplankton as a result of the discretness effect and change in pigment composition,' IAMAP Scientific Assembly (Hamburg), (extended abstract) 18-20.

Bricaud, A., Morel, A., and Prieur, L., 1981. 'Absorption of dissolved organic matter of the sea ('yellow substance') in the uv and visible domains,' Limnology and Oceanography, 26, 43-53.

Clark, D.K., Baker, E.T., and Strong, A.E., 1980. 'Upwelled spectral radiance distribution in relation to particulate matter in sea water,' Boundary Layer Meteorology, 18, 287-298.

Clark, D.K.,, 1981. 'Phytoplankton algorithms for the Nimbus-7 CZCS,' in 'Oceanography from Space,' edited by J.R.F. Gower, Plenum Press, New York, pp. 227-238.

Clarke, G.K., Ewing, G.C., and Lorenzen, C.J., 1970. 'Spectra of backscattered light from the sea obtained from aircraft as a measure of chlorophyll concentration,' Science, 167, 1119-1121.

Clarke, G.K., and Ewing, G.C., 1974. 'Remote spectroscopy of the sea for biological production studies,' in Optical Aspects of Oceanography, edited by N.G. Jerlov and E.S. Nielsen, Academic Press, London. Ch. XVII, 389-413.

Duntley, S.Q., 1942. 'Optical properties of diffusing materials,' J. Opt. Soc. Am., 32, 61-70.

Gordon, H.R., 1976. 'Radiative transfer: a technique for simulating the ocean in satellite remote sensing calculations,' Applied Optics, 15, 1974-1979.

Gordon, H.R., 1978. 'Removal of atmospheric effects from satellite imagery of the oceans,' Applied Optics, 17, 1631-1636.

Gordon, H.R., 1981a. 'A preliminary assessment of the Nimbus-7 CZCS atmospheric correction algorithm in a horizontally inhomogeneous atmosphere,' in 'Oceanography from Space,' edited by J.R.F. Gower, Plenum Press, New York, pp. 257-266.

Gordon, H.R., 1981b. 'Reduction of error introduced in the processing of coastal zone color scanner-type imagery resulting from sensor calibration and solar irradiance uncertainty,' Applied Optics, 20, 207-210.

Gordon, H.R., 1981c. 'Remote sensing of ocean properties at optical wavelengths,' IAMAP Scientific Assembly (Hamburg), (extended abstract) 128-131.

Gordon, H.R., Brown, O.B., and Jacobs, M.M., 1975. 'Computed relationships between the inherent and apparent optical properties of a flat homogeneous ocean,' Applied Optics, 14, 417-427.

Gordon, H.R., and Clark, D.K., 1980a. 'Remote sensing optical properties of a stratified ocean: an improved interpretation,' Applied Optics, 19, 3428-3430.

Gordon, H.R., and Clark, D.K., 1980b. 'Atmospheric effects in the remote sensing of phytoplankton pigments,' Boundary Layer Meteorology, 18, 299-313.

Gordon, H.R., and Clark, D.K., 1981. 'Clear water radiances for atmospheric correction of coastal zone color scanner imagery,' Applied Optics, 20, 4175-4180.

Gordon, H.R., Clark, D.K., Brown, J.W., Brown, O.B., and Evans, R.H., 1982. 'Satellite measurement of the phytoplankton pigment concentration in the surface waters of a warm core Gulf Stream ring,' J. Mar. Res., 40, 491-502.

Gordon, H.R., Clark, D.K., Brown, J.W., Brown, O.B., Evans, R.H., and Broenkow, W.W., 1983. 'Phytoplankton pigment concentrations in the Middle Atlantic Bight: comparison between ship determinations and Coastal Zone Color Scanner estimates,' Applied Optics, 22, 20–36.

Gordon, H.R., Clark, D.K., Mueller, J.L., and Hovis, W.A., 1980. 'Phytoplankton pigments derived from the Nimbus-7 CZCS: initial comparisons with surface measurements,' Science, 210, 63–66.

Gordon, H.R., and McCluney, W.R., 1975. 'Estimation of the depth of sunlight penetration in the sea for remote sensing,' Applied Optics, 14, 413–416.

Gordon, H.R., Mueller, J.L., and Wrigley, R.C., 1979. 'Atmospheric correction of Nimbus-7 coastal zone color scanner imagery,' Presented at IFAORS Workshop on 'Interpretation of Remotely Sensed Data,' Williamsburg, Virginia, May 23–25 (also in 'Remote Sensing of Oceans and Atmospheres,' edited by A. Deepak, Academic Press, New York, 1980).

Gower, J.F.R., Denman, K.L., and Holyer, R.J., 1980. 'Phytoplankton patchiness indicates the fluctuation spectrum of mesoscale oceanic turbulence,' Nature, 288, 157–159.

Gower, J.F.R., and Denman, K.L., 1981. 'Reply to Satellite sensed turbulent ocean structure,' Nature, 294, 693–694.

Hobson, L.A., Menzel, D.W., and Barber, R.T., 1973. 'Primary productivity and sizes of pools of organic carbon in the mixed layer of the ocean,' Mar. Biol., 19, 298–306.

Højerslev, N. and Jerlov, N.G., 1977. 'The use of the colour index for determining quanta irradiance in the sea,' Rep. Inst. Phys. Oceanogr., Univ. Copenhagen, No. 35, 12pp.

Højerslev, N., 1980. 'Water colour and its relation to primary production,' Boundary Layer Meteorology, 18, 203–220.

Højerslev, N., 1981. 'Assessment of some suggested algorithms on sea colour and surface chlorophyll,' in 'Oceanography from Space,' edited by J.R.F. Gower, Plenum Press, New York, p. 347–354.

Hovis, W.A., 1981. 'The Nimbus-7 coastal zone color scanner (CZCS) program,' in 'Oceanography from Space,' edited by J.R.F. Gower, Plenum Press, New York, pp. 213–225.

Hovis, W.A., and Leung, K.C., 1977. 'Remote sensing of ocean color,' Optical Engineering, 16, 153–166.

Hovis, W.A., Clark, D.K., Anderson, F.,Austin, R.W. Wilson, W.H., Baker, E.T., Ball, D., Gordon, H.R., Mueller, J.L., El Sayed, S.Y., Sturm, B., Wrigley, R.C., and Yentsch, C.S., 1980. 'Nimbus-7 coastal zone color scanner: system description and initial imagery,' Science, 210, 60–63.

Innamorati, M., 1978. 'Spettri della radiazione sottemarina nell'arcipelago delle Galapagos,' in Galapagos, studi e ricerche, Gruppo di Richerche scientifiche e techniche., Florence, 1–59.

Jain, S.C., and Miller, J.R.,, 1976. 'Subsurface water parameters: optimization approach to their determination from remotely sensed water color data,' Applied Optics, 15, 886–890.

Jerlov, N.G., 1974. 'Significant relationships between optical properties of the sea,' in 'Optical Aspects of Oceanography,' edited by N.G. Jerlov and E.S. Nielsen, Academic Press, London. Ch. IV, 77–94.

Joseph, J., 1950. 'Untersuchungen über Ober- und Unterlichtmessungen in Meere und über ihren Zusammenhang mit Durchsichtigkeits messungen,' Deut. Hydrograph., 3, 324–335.

Kirk, J.T.O., 1976. 'Yellow substance (Gelbstoff) and its contribution to the attenuation of photosynthetically active radiation in some inland and coastal southeastern Australian waters,' Aust. J. Mar. Freshwater Res., 27, 61-71.

Kirk, J.T.O., 1981. 'Monte Carlo study of the nature of the underwater light field in, and relationships between optical properties of, turbid yellow waters,' Aust. J. Mar. Freshwater Res., 32, 517-532.

Kozlyaninov, M.V., 1972. 'The basic relationships between the hydro-optical parameters,' in 'Optics of the Ocean and the Atmosphere,' edited by K.S. Shifrin, Nauka, pp. 5-24 (in Russian).

Kozlyaninov, M.V. and Pelevin, V.N., 1965. 'On the application of a one-dimensional approximation in the investigation of the propagation of optical radiation in the sea,' Tr. Inst. Okeanol. Akad. Nauk. SSSR, 77, 73-79. Also, 1966, U.S. Dept. Comm. Jt. Publ. Res. Ser. Rep., 36, (816) 54-63 (English translation).

Lesieur, M., and Sadourny, R., 1981. 'Satellite-sensed turbulent ocean structure,' Nature, 294, 674.

Maul, G.A., and Gordon, H.R., 1975. 'On the use of the Earth Resources Technology Satellite (LANDSAT-I) in optical oceanography,' Rem. Sens. Environ., 4, 95-128.

Morel, A., 1970. 'Examen des resultats experimentaux concernant la diffusion de la lumiere par les eaux de mer,' in 'Electromagnetics of the Sea,' AGARD Conference Proceedings, 77, 300-309.

Morel, A., 1973a. 'Measurements of spectral and total radiant flux,' p. F1-F341, in SCOR-UNESCO Data Rep. Discoverer Expedition, edited by J.E. Tyler, S.I.O. Ref. 73-16.

Morel, A., 1973b. 'Diffusion de la lumiere par les eaux de mer; resultats experimentaux et approche theorique,' in 'Optics of the Sea,' AGARD Lecture Series, 63, Sect. 3, 1-76.

Morel, A., 1974. 'Optical properties of pure water and sea water,' in 'Optical Aspects of Oceanography,' edited by N.G. Jerlov and E.S. Nielsen, Academic Press, London. Ch. I, 1-24.

Morel, A., 1978. 'Available, usable, and stored radiant energy in relation to marine photosynthesis,' Deep Sea Res., 25, 673-688.

Morel, A., 1979. 'Depth of the euphotic zone, average pigment concentration, and primary production efficiency,' IAPSO-UGGI XVII General Assembly (Canberra), Proces-verbaux, 15, 116-117.

Morel, A., 1980. 'In-water and remote measurement of ocean color,' Boundary Layer Meteorology, 18, 177-201.

Morel, A., 1982. 'Optical properties of radiant energy in the waters of the Guinea dome and the Mauritanian upwelling area in relation to primary production,' Rapp. P-v. Reun. Cons. Int. Explor. Mer., 180, 94-107.

Morel, A., and Bricaud A., 1981a. 'Theoretical results concerning the optics of phytoplankton, with special reference to remote sensing applications,' in 'Oceanography from Space,' edited by J.R.F. Gower, Plenum Press, New York, pp. 313-328.

Morel, A., and Bricaud A., 1981b. 'Theoretical results concerning light absorption in a discrete medium and application to the specific absorption of phytoplankton,' Deep Sea Res., 28A, 11, 1357-1393.

Morel, A., and Caloumenos, L., 1973. 'Mesure d'eclairements sous marins, flux de photons et analyse spectrale,' Centre Rech. Oceanogr., Villefranche-sur-mer, Rapp. 11, pp242.

Morel, A., and Gordon, H.R.,, 1980. 'Report of the working group on water color,' Boundary Layer Meteorology, 18, 343-355.

Morel, A., and Prieur, L., 1975a. 'Analyse spectrale des coefficients d'attenuation diffuse, de reflexion diffuse, d'absorption, et de retrodiffusion pour diverses regions marines,' Centre Rech. Oceanogr., Villefranche-sur-mer Rapp. 17, 157pp.

Morel, A., and Prieur, L., 1975b. 'Analyse spectrale du facteur de reflexion diffuse de la mer. IAPSO-IGGU XVI General Assembly (Grenoble), Proces-verbaux, 14, 177-178.

Morel, A., and Prieur, L., 1976. 'Eclairements sous marins,' in Resultats des Campagnes a la mer, No 10, CINECA 5-Charcot, 1-256. Publications CNEXO.

Morel, A., and Prieur, L., 1977a. 'Analysis of variations in ocean color,' Limnology and Oceanography, 22, 709-722.

Morel, A., and Prieur, L., 1977b. 'Energie radiative disponible pour la photosynthese,' Resultats des campagnes a la mer, No 13, fasc. 2, Campagne GUIDOM-Charcot, 33-62, Publications CNEXO.

Morel, A., Prieur, L., and Matsumoto, M., 1978. 'Mesures d'optique marine,' Resultats des campagnes a la mer, No 6, Campagne ANTIPROD 1, 99-141, Publications CNEXO.

Morel, A., and Smith, R.C., 1982. Terminology and units in optical oceanography, Marine Geodesy, 5, 335-349.

Mueller, J.L., and LaViolette, P.E., 1981. 'Signatures of ocean fronts observed with the Nimbus-7 CZCS,' in 'Oceanography from Space,' edited by J.R.F. Gower, Plenum Press, New York, pp. 295-302.

Neckel, H., and Labs, D., 1981. 'Improved data of solar spectral irradiance from 0.33 to 1.25μ,' Solar Physics, 74, 231–249.

Okami, N., Kishino, M., and Sugihara, S., 1978. 'Measurements of spectral irradiance in the seas around the Japanese Islands,' Tech Rep. of Phys. Oceanogr. Lab., No 2, 129pp.

Okami, N., Kishino, M., Sugihara, S., Unoki, S., Muneyama, K., Toyota, T., Nakajima, T., Sasaki, Y., and Emura, T., 1981 'Measurements of spectral irradiance in Tokyo Bay,' Tech Rep. of Phys. Oceanogr. Lab., No 5, 75pp.

Parsons, T.R., and Takahashi, M., 1973. 'Biological Oceanographic Processes,' Pergamon Press, Oxford. 184pp.

Platt, T., Denman, K.L., and Jassby, A.D., 1977. 'Modeling the productivity of phytoplankton,' in 'The Sea,' Vol. 6, edited by Goldberg, E.D., McCave, I.N., O'Brien, J.J., and Steele, J.H., Ch. 21, 807–856.

Preisendorfer, R.W., 1961. 'Application of radiative transfer theory to light measurements in the sea,' UGGI Monogr. No.10 (Symposium on Radiant Energy in the Sea), 11–30

Prieur, L., 1976. 'Transfert radiatif dans les eaux de mer. Application a la determination de parametres optiques caracterisant leur teneur en substances dissoutes et leur contenu en particules,' D.Sci. Thesis, Univ. Pierre et Marie Curie, 243pp.

Prieur, L., and Morel, A., 1975. 'Relations theoriques entre le facteur de reflexion diffuse de l'eau de mer a diverses profondeurs et les caracteristiques optiques (absorption, diffusion),' IAPSO–IGGU XVI General Assembly (Grenoble),

Prieur, L., and Sathyendranath, S., 1981. 'An optical classification of coastal and oceanic waters based on the specific spectral absorption curves of phytoplankton pigments, dissolved organic matter, and other particulate materials,' Limnology and Oceanography., 26, 671-689.

Quenzel, H., and Kaestner, M., 1980. 'Optical properties of the atmosphere: calculated variability and application to satellite remote sensing of phytoplankton,' Applied Optics, 19, 1338-1344.

Ramsey, R.C., and White, P.G., 1973. 'Ocean color data analysis applied to MOCS and SIS data,' Final Report. NOAA Contract No. N62306-72-C-0037, 75pp.

Smith, R.C., 1973. 'Scripps spectroradiometer data,' pp. G1-G160, in SCOR-UNESCO Data Rep. Discoverer Expedition, edited by J.E. Tyler, S.I.O. Ref. 73-16.

Smith, R.C., 1974. 'Structure of solar radiation in the upper layers of the sea,'in 'Optical Aspects of Oceanography,' edited by N.G. Jerlov and E.S. Nielsen, Academic Press, London. Ch. V, 95-119.

Smith, R.C., and Baker, K.S., 1978a. 'The bio-optical state of ocean waters and remote sensing,' Limnology and Oceanography, 23, 247-259.

Smith, R.C., and Baker, K.S., 1978b. 'Optical classification of natural waters,' Limnology and Oceanography, 23, 260-267.

Smith, R.C., and Baker, K.S., 1982. 'Oceanic chlorophyll concentrations as determined using Nimbus-7 Coastal Zone Color Scanner imagery,' J. Mar. Biol., 66, 269-279.

Smith, R.C., Eppley, R.W., and Baker, K.S., 1982. 'Application of satellite CZCS chlorophyll images for the study of primary production in Southern California coastal waters', J. Mar. Biol., 66, 281-288.

Smith, R.C., and Wilson, W.H., 1981. 'Ship and satellite bio-optical research in the California Bight,' in 'Oceanography from Space,' edited by J.R.F. Gower, Plenum Press, New York, p. 281-294.

Sturm, B., 1981. 'Ocean color remote sensing and the retrieval of surface chlorophyll in coastal waters using the Nimbus-7 CZCS,' in 'Oceanography from Space,' edited by J.R.F. Gower, Plenum Press, New York, p. 267-280.

Tanre, D., Herman, M., Deschamps, P.Y., and de Leffe, A., 1979. 'Atmospheric modeling for space measurements of ground reflectances, including bidirectional properties,' Applied Optics, 18, 3587-3594.

Tassan, S., 1981. 'A global sensitivity analysis for the retrieval of chlorophyll concentrations from remotely sensed radiances – the influence of wind,' in 'Oceanography from Space,' edited by J.R.F. Gower, Plenum Press, New York, p. 371-376.

Tyler, J.E., 1960. 'Radiance distribution as a function of depth in an underwater environment,'Bulletin of the Scripps Institution of Oceanography of the University of California, La Jolla, California, 7, 363-412.

Tyler, J.E., and Smith, R.C., 1970. 'Measurements of Spectral Irradiance,' Gordon and Breach, New York, 103pp.

Viollier, M., Deschamps, P.Y., Lecomte, P., 1978. 'Airborne remote sensing of chlorophyll content under cloudy sky as applied to the tropical waters in the Gulf of Guinea,' Remote Sensing of Environment, 7, 235-248.

Viollier, M.,Tanre, D., and Deschamps, P.Y., 1980. 'An algorithm for remote sensing of water color from space,' Boundary Layer Meteorology, 18, 247-267.

Viollier, M., 1982.'Radiance calibration of the
Coastal Zone Color Scanner: a proposed adjustment,'
Applied Optics, 21, 1142–1145.

Wrigley, R.C., 1980. 'Frontal activity in Northern
Central Pacific via CZCS,' Trans. Amer. Geophys.
Union, 46, 1001.